Theoretical Fluid Mechanics

Raj Spielmann

Theoretical Fluid Mechanics

An application-oriented introduction with detailed derivations

 Springer

Raj Spielmann
Muri bei Bern, Switzerland

ISBN 978-3-662-72239-8   ISBN 978-3-662-72240-4   (eBook)
https://doi.org/10.1007/978-3-662-72240-4

This book is a translation of the original German edition "Theoretische Strömungsmechanik" by Raj Spielmann, published by Springer-Verlag GmbH, DE in 2025. The translation was done with the help of an artificial intelligence machine translation tool. A subsequent human revision was done primarily in terms of content, so that the book will read stylistically differently from a conventional translation. Springer Nature works continuously to further the development of tools for the production of books and on the related technologies to support the authors.

Translation from the German language edition: "Theoretische Strömungsmechanik" by Raj Spielmann, © Der/die Herausgeber bzw. der/die Autor(en), exklusiv lizenziert an Springer-Verlag GmbH, DE, ein Teil von Springer Nature 2025. Published by Springer Berlin Heidelberg. All Rights Reserved.

Editorial Contact: Veronika Erdmann
This Springer imprint is published by the registered company Springer-Verlag GmbH, DE, part of Springer Nature.
The registered company address is: Heidelberger Platz 3, 14197 Berlin, Germany

If disposing of this product, please recycle the paper.

For Olja, who made it possible for me to write this book

Preface

Among graduates of scientific disciplines, particularly mathematicians, only a relatively small proportion go into applied research, which is traditionally dominated by engineers. One of the reasons is the gap between theoretically oriented education and the demands of practice. Like other theories that have developed over centuries, fluid mechanics also contains a wealth of important details that can hardly be accommodated in a short textbook. When choosing the necessary abridgements, each specialist is guided by their target audience. Mathematicians tend to shorten or omit the discussion of applications, while engineers shorten certain mathematical details on the way to the results, especially as they can be compensated for by computer calculations. With this compromise, there is a risk that some aspects will be lost from view.

This book aims to help overcome this barrier. It is primarily aimed at students and lecturers of university colleges in the mathematical-scientific field. At the same time, it should be useful for engineers who are interested in the derivation of the formulas and backgrounds. Unlike other modern presentations that assume greater familiarity with the subject and its methods, it gently guides the beginner through the theory to practical examples.

Definitions and central concepts are highlighted in the text. To motivate the theory, the applications covered in the corresponding sections are briefly introduced at the beginning and discussed in detail after the formulas are provided.

An introductory chapter is used to practice the necessary calculation techniques that are only marginally covered in the basic mathematics lectures. If the derivations are rarely found in standard textbooks, they are given here. Otherwise we refer to them.

The main part begins with the derivation of the basic equations of fluid mechanics. The explanation of aerodynamic lift is very extensive and is only touched upon in introductory texts. Therefore, it is in the foreground here. After a discussion of the boundary layer as the main location of vortex formation, we present Prandl's three-dimensional lift theory. This allows the relevant forces for specific wing profiles to be calculated in the low flight speed range. Finally, we present some applications from hydrodynamics that show different facets of the Navier-Stokes model and are characterized by particular clarity.

As a mathematical background, basic knowledge from four semesters of analysis is required. To appeal to a broader audience, numerous small and sometimes tiring calculation steps are not left to the reader, but formulated as problems that are fully recalculated in the solution part. Thus, the more experienced reader can follow the presentation in a condensed form, while the beginner continues reading after a glance at the solution part. Details of mathematical derivations, which are used less frequently and exceed the scope of the main text, are moved to the appendix "Formulas, Tables and Supplements."

To simplify readability, the concrete enumeration of mathematical prerequisites such as smoothness of curves and areas, differentiability and integrability of functions is largely dispensed with. As far as possible, we assume that the corresponding requirements are met. These details are indispensable for mathematical existence and uniqueness proofs, while they are less often needed for understanding physical mechanisms. A prominent exception is the trailing edge of an aerofoil, which is of essential importance for aerodynamic lift.

To avoid erroneous copyright infringements, I have decided to create most of the illustrations myself. For this reason, the book contains some pencil sketches instead of photos.

The manuscript received significant corrections from Prof. Thomas P. Witelski and additions from Prof. Maximilian Platzer, who pointed out important aspects of aerodynamic lift. Excellent cooperation with Springer Spectrum was ensured by the editors Veronika Erdmann and Iris Ruhmann, whose criticism contributed to a better understanding of the text. However, my special thanks go to two people. First to my wife Olga, who patiently and persistently stood by my side during long work phases. I am also deeply indebted to Prof. em. Albert Fässler. He not only provided valuable suggestions but also gave me the strength and support to complete the project at a critical moment.

Muri bei Bern, Switzerland   Raj Spielmann
July 2025

Declarations

Competing Interests The author has no competing interests to declare that are relevant to the content of this manuscript.

Contents

Mathematical Tools

1

Aerodynamic calculations require a strong familiarity with vector analysis. The formulas presented here are used regularly throughout the text. The same applies to the problems, the solutions to which can be found at the end of the book.

1.1   Vectors

In the following, scalars are written in normal print and vectors in bold print. Thus, $\mathbf{x}$ denotes a point in space, $p = p(\mathbf{x})$ the pressure present there and $\mathbf{u} = \mathbf{u}(\mathbf{x})$ the velocity. Magnitudes of vectors appear in double vertical bars or are set in normal print, for example $r = \|\mathbf{r}\|$.

Partial derivatives are abbreviated as follows:

$$\partial_x p = \frac{\partial p}{\partial x}, \quad \partial_y p = \frac{\partial p}{\partial y}, \quad \partial_z p = \frac{\partial p}{\partial z}$$

The calculations are simplified by a shortened notation.

- For *vector fields* $\mathbf{a} = (a_x, a_y, a_z)$, we write for short $\mathbf{a} = (a_i)$, while $\nabla \mathbf{a} = (\partial_j a_i)$ denotes the associated *Jacobian matrix*.

  The term a_i (without brackets) merely indicates the i-th component of $\mathbf{a}$.

  In most intermediate calculations, we allow ourselves the inaccuracy of omitting the above vector or tensor brackets. Long terms thus appear less bulky, which benefits their readability.
- According to Einstein's summation convention, we sum over double indices.
- It is convenient to use the *Kronecker symbol*:

$$\delta_{ij} = \begin{cases} 1 & \text{for } i = j, \\ 0 & \text{for } i \neq j. \end{cases} \tag{1.1}$$

© The Author(s), under exclusive license to Springer-Verlag GmbH, DE, part of Springer Nature 2026
R. Spielmann, *Theoretical Fluid Mechanics*,
https://doi.org/10.1007/978-3-662-72240-4_1

In particular, we obtain in classical and shortened notation

Scalar product
$$\mathbf{a} \cdot \mathbf{b} = a_x b_x + a_y b_y + a_z b_z \tag{1.2}$$
$$= a_i b_i$$

Vector product
$$\mathbf{a} \times \mathbf{b} = \epsilon_{ijk} a_i b_j \mathbf{e}_k \tag{1.3}$$
$$= (\epsilon_{ijk} a_i b_j)$$

In the latter, ϵ_{ijk} denotes the *Levi-Civita symbol*:

$$\epsilon_{ijk} = \begin{cases} +1 & \text{for a permutation of even order} \\ -1 & \text{for a permutation of odd order} \\ 0 & \text{if at least two indices are equal} \end{cases} \tag{1.4}$$

It is calculated using the rule

$$\epsilon_{ijk} = \frac{1}{2}(i - j)(j - k)(k - i) \tag{1.5}$$

Problem 1.1 Prove the summation rule

$$\epsilon_{ijk}\epsilon_{imn} = \delta_{jm}\delta_{kn} - \delta_{jn}\delta_{km} \tag{1.6}$$

A simple way to memorize the vector product is the following determinant.

$$\mathbf{a} \times \mathbf{b} = \det \begin{bmatrix} \mathbf{e}_x & \mathbf{e}_y & \mathbf{e}_z \\ a_x & a_y & a_z \\ b_x & b_y & b_z \end{bmatrix}$$

The following proves useful for calculating volumes.

Box product
$$\mathbf{a} \cdot (\mathbf{b} \times \mathbf{c}) = \det \begin{bmatrix} a_x & a_y & a_z \\ b_x & b_y & b_z \\ c_x & c_y & c_z \end{bmatrix} \tag{1.7}$$
$$= \epsilon_{ijk} a_i b_j c_k$$

From the formula, it is evident that any odd permutation of $\mathbf{a}, \mathbf{b}, \mathbf{c}$ changes the sign of the box product, while it remains unchanged for even permutations.

Problem 1.2 Prove in (1.7) that the addition of a multiple of one row to another does not change the value of the determinant.

Problem 1.3 Show the rule

$$\det \begin{bmatrix} ka_x & ka_y & ka_z \\ b_x & b_y & b_z \\ c_x & c_y & c_z \end{bmatrix} = k\det \begin{bmatrix} a_x & a_y & a_z \\ b_x & b_y & b_z \\ c_x & c_y & c_z \end{bmatrix}$$

For vector fields $\mathbf{a} = (a_x, a_y, a_z)$ and scalar fields b, we use the following operators, whose notation is first given classically and then in shortened notation.

Curl

$$\operatorname{rot}\mathbf{a} = \epsilon_{ijk}\partial_i a_j \mathbf{e}_k \tag{1.8}$$

$$= (\epsilon_{ijk}\partial_i a_j)$$

In analogy to the vector product, we write it in the form

$$\operatorname{rot}\mathbf{a} = \det \begin{bmatrix} \mathbf{e}_x & \mathbf{e}_y & \mathbf{e}_z \\ \partial_x & \partial_y & \partial_z \\ a_x & a_y & a_z \end{bmatrix}$$

Divergence

$$\operatorname{div}\mathbf{a} = \partial_x a_x + \partial_y a_y + \partial_z a_z \tag{1.9}$$

$$= \partial_i a_i$$

Laplace Operator

$$\Delta\mathbf{a} = \Delta a_k \mathbf{e}_k = \begin{pmatrix} \Delta a_x \\ \Delta a_y \\ \Delta a_z \end{pmatrix} \tag{1.10}$$

$$= (\partial_i \partial_i a_j)$$

Gradient

$$\operatorname{grad} b = \partial_k b\, \mathbf{e}_k = \begin{pmatrix} \partial_x b \\ \partial_y b \\ \partial_z b \end{pmatrix} \tag{1.11}$$

$$= (\partial_i b)$$

The *directional derivative* of the vector field $\mathbf{b}$ along $\mathbf{a}$ is defined by

$$(\mathbf{a}\cdot\nabla)\,\mathbf{b} = \begin{pmatrix} a_x\partial_x b_x + a_y\partial_y b_x + a_z\partial_z b_x \\ a_x\partial_x b_y + a_y\partial_y b_y + a_z\partial_z b_y \\ a_x\partial_x b_z + a_y\partial_y b_z + a_z\partial_z b_z \end{pmatrix} = (a_i\partial_i b_j) \tag{1.12}$$

Problem 1.4 Prove the identities

$$\operatorname{rot}(b\,\mathbf{a}) = (\operatorname{grad} b)\times\mathbf{a} + b\operatorname{rot}\mathbf{a} \tag{1.13}$$

$$\text{rot}\,(\text{rot}\,\mathbf{a}) = \text{grad}\,\text{div}\,\mathbf{a} - \Delta\mathbf{a} \tag{1.14}$$

$$\text{div}\,(b\,\mathbf{a}) = b\,\text{div}\,\mathbf{a} + \mathbf{a}\,\text{grad}\,b \tag{1.15}$$

$$\text{rot}\,(\mathbf{a} \times \mathbf{b}) = (\mathbf{b} \cdot \nabla)\mathbf{a} - (\mathbf{a} \cdot \nabla)\mathbf{b} + \mathbf{a}\,(\text{div}\,\mathbf{b}) - \mathbf{b}\,(\text{div}\,\mathbf{a}) \tag{1.16}$$

$$\text{grad}\,(\mathbf{a} \cdot \mathbf{b}) = (\mathbf{a} \cdot \nabla)\,\mathbf{b} + (\mathbf{b} \cdot \nabla)\,\mathbf{a} + \mathbf{a} \times \text{rot}\,\mathbf{b} + \mathbf{b} \times \text{rot}\,\mathbf{a} \tag{1.17}$$

$$\text{div}\,\text{rot}\,\mathbf{a} = 0 \tag{1.18}$$

$$\text{rot}\,\text{grad}\,b = 0 \tag{1.19}$$

A *planar vector field* is a vector field in the xy-plane.

Problem 1.5 Prove the formula

$$\text{rot}\,\mathbf{u} = \begin{pmatrix} 0 \\ 0 \\ \partial_x v - \partial_y u \end{pmatrix} \qquad \text{for } \mathbf{u} = \begin{pmatrix} u(x,\,y) \\ v(x,\,y) \\ 0 \end{pmatrix} \tag{1.20}$$

Since rot $\mathbf{u}$ is perpendicular to the xy-plane, it is described by a single component and can be identified with a scalar field. This allows the following short forms.

A planar vector field is denoted by $\mathbf{u} = \begin{pmatrix} u(x,\,y) \\ v(x,\,y) \end{pmatrix}$. The notation

$$\text{div}\,\mathbf{u} = \partial_x u + \partial_y v, \qquad \text{grad}\,\mathbf{u} = \begin{pmatrix} \partial_x u \\ \partial_y v \end{pmatrix}, \qquad \text{rot}\,\mathbf{u} = \partial_x v - \partial_y u \tag{1.21}$$

is common here.

Planar vector fields are suitable for modeling if the flow is completely or approximately in parallel planes. The corresponding obstacles must have a cylindrical shape. We can imagine them either as an infinitely long or as a finite cylinder between two parallel plates on which the fluid moves without friction. The two-dimensional flow is the flow on any plane parallel to the plates. Even in the two-dimensional representation, the obstacle is called a cylinder (see Fig. 1.1).

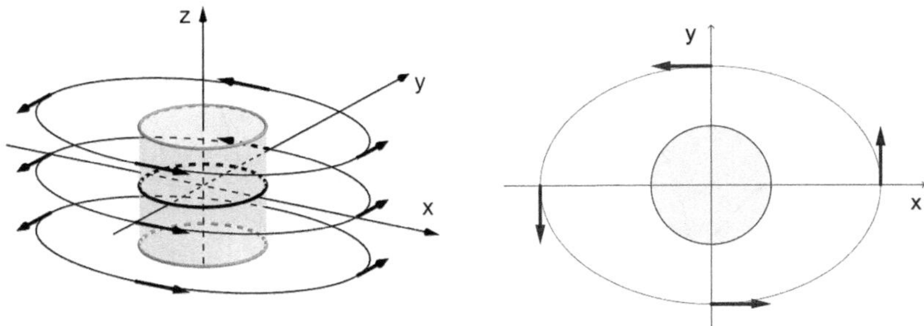

Fig. 1.1 (left) Flow around a circular cylinder. (right) Corresponding two-dimensional representation.

1.2   Integrals

In this section, $\mathbf{u} = \mathbf{u}(\mathbf{x}, t)$ denotes a vector field and $\mathsf{T} = \mathsf{T}(\mathbf{x}, t)$ a tensor field, where the existence of the corresponding derivatives and the integrability with respect to $\mathbf{x}$ are implicitly assumed. Here, the time t is treated as a parameter. Time derivatives are denoted by $\partial/\partial t$ or ∂_t, while the notation ∂_i is only used for spatial derivatives.

For a *second-order tensor* $\mathsf{T} = (T_{ij})$, the *divergence* is defined by

$$\operatorname{div} \mathsf{T} = (\partial_j T_{ij})$$

and thus represents a vector, see [1].

The product $\mathsf{T} \cdot \mathbf{n}$ of a second-order tensor with a vector $\mathbf{n}$ again forms a vector.

Integration on boundaries and enclosed domains is only possible if they have an appropriate structure.

Let $C : [a, b] \to \mathbb{R}^3$ be a continuous curve. It is said to be
- *simple* if $C(t) \neq C(t')$ for $t \neq t'$
- *closed* if $C(a) = C(b)$
- *simple closed* if it is closed and simple on $(a, b]$.

Consequently, a simple closed curve has no self-intersections (see Fig. 1.2).

Fig. 1.2 (left) Simple close curve. (right) Self-intersecting curve.

Theorem 1.1 (Gauss's Integral Theorem)

Two-dimensional case: Let $A \subset \mathbb{R}^2$ be a domain having a simple closed, piecewise smooth [1] boundary curve $S = \partial A$ and outward pointing unit normal $\mathbf{n}$. Further, let $\mathbf{u}$ be a planar vector field. Then

$$\int_A \operatorname{div} \mathbf{u} \, dA = \int_{\partial A} \mathbf{u} \cdot \mathbf{n} \, ds \tag{1.22}$$

Three-dimensional case: Let $V \subset \mathbb{R}^3$ be a domain having a smooth closed surface $S = \partial V$ and outward pointing unit normal $\mathbf{n}$. Further, let $\mathbf{u}$ be a vector field and T a second-order tensor. Then

$$\int_V \operatorname{div} \mathbf{u} \, dV = \int_{\partial V} \mathbf{u} \cdot \mathbf{n} \, dS$$

$$\int_V \operatorname{div} \mathsf{T} \, dV = \int_{\partial V} \mathsf{T} \cdot \mathbf{n} \, dS$$

Proof For planar as well as three-dimensional vector fields $\mathbf{u}$, the derivation can be found in standard textbooks on multidimensional analysis. What remains is the proof in the three-dimensional case, where $\mathsf{T} = (T_{ij})$ is a tensor field. Let $\mathbf{a} = (a_i)$ be a constant vector. For the unit normal on the surface ∂V, we write $\mathbf{n} \cdot dS = (dS_j)$. Using Einstein's summation convention, we obtain

$$\mathbf{a} \cdot \int_V \operatorname{div} \mathsf{T} \, dV = a_i \int_V \partial_j T_{ij} \, dV = \int_V \partial_j (a_i \, T_{ij}) \, dV$$

---

[1] A piecewise smooth curve consists of a finite number of segments, each of which is differentiable and has continuous derivatives. This includes boundaries of airfoil profiles with a sharp trailing edge.

The last term contains the vector $(a_i T_{ij})$, so we can apply Gauss's theorem in vector form. Then,

$$\int_V \partial_j (a_i T_{ij})\, dV = \int_{\partial V} (a_i T_{ij})\, dS_j = a_i \int_{\partial V} T_{ij}\, dS_j = \mathbf{a} \cdot \int_{\partial V} \mathsf{T} \cdot \mathbf{n}\, dS,$$

and consequently

$$\mathbf{a} \cdot \int_V \operatorname{div} \mathsf{T}\, dV = \mathbf{a} \cdot \int_{\partial V} \mathsf{T} \cdot \mathbf{n}\, dS$$

Since $\mathbf{a}$ is arbitrary, we have verified our assertion. $\qquad\square$

Problem 1.6 Show for planar vector fields $\mathbf{u} = \begin{pmatrix} u \\ v \end{pmatrix}$ that the differential on the right-hand side of (1.22) can be chosen in Cartesian coordinates in the form

$$\mathbf{n}\, ds = \begin{pmatrix} dy \\ -dx \end{pmatrix} \qquad \text{or} \qquad \mathbf{n}\, ds = \begin{pmatrix} -dy \\ dx \end{pmatrix}$$

and derive the following coordinate formula from (1.22).

$$\int_A \left(\frac{\partial u}{\partial x} + \frac{\partial v}{\partial y} \right) dxdy = \int_{\partial A} (u\, dy - v\, dx) \tag{1.23}$$

Problem 1.7 Let $\Sigma \subset \mathbb{R}^3$ be the surface of a sphere. Prove that

$$\int_\Sigma \mathbf{n}\, dS = 0$$

Corollary 1.1 *Let $V \subset \mathbb{R}^3$ be a domain having a smooth closed surface $S = \partial V$ and outward pointing unit normal $\mathbf{n}$. Further, let $\mathbf{a}$ and $\mathbf{u}$ be vector fields. Then*

$$\int_{\partial V} \mathbf{a}(\mathbf{n} \cdot \mathbf{u})\, dS = \int_V [\mathbf{a}(\operatorname{div} \mathbf{u}) + (\mathbf{u} \cdot \nabla)\mathbf{a}]\, dV \tag{1.24}$$

Proof For $\mathbf{a} = (a_i)$ and $\mathbf{u} = (u_j)$ we define the tensor T by $T_{ij} = a_i u_j$. Then,

$$\int_V \partial_j (a_i u_j)\, dV = \int_{\partial V} a_i u_j\, dS_j \qquad \text{using Theorem 1.1}$$

$$\int_V [u_j\, \partial_j a_i + (\partial_j u_j) a_i]\, dV = \int_{\partial V} \mathbf{a}(\mathbf{u} \cdot \mathbf{n})\, dS$$

$$\int_V [(\mathbf{u} \cdot \nabla)\mathbf{a} + (\operatorname{div} \mathbf{u})\mathbf{a}]\, dV = \int_{\partial V} \mathbf{a}(\mathbf{u} \cdot \mathbf{n})\, dS \qquad \text{on the left by (1.9), (1.12)}$$

$$\square$$

Corollary 1.2 *Let $V \subset \mathbb{R}^3$ be a domain having a smooth closed surface $S = \partial V$ and outward pointing unit normal $\mathbf{n}$. Further, let f be a scalar function. Then*

$$\int_V \operatorname{grad} f \, \mathrm{d}V = \int_{\partial V} f \mathbf{n} \, \mathrm{d}S \tag{1.25}$$

Proof Let $\mathbf{a}$ be a constant vector. Then,

$$\int_V \operatorname{div}(f\mathbf{a}) \, \mathrm{d}V = \int_{\partial V} f\mathbf{a} \cdot \mathbf{n} \, \mathrm{d}S \qquad \text{using Theorem 1.1}$$

$$\int_V (f \operatorname{div}\mathbf{a} + \mathbf{a} \cdot \operatorname{grad} f) \, \mathrm{d}V = \int_{\partial V} f\mathbf{a} \cdot \mathbf{n} \, \mathrm{d}S \qquad \text{by (1.15)}$$

$$\int_V \mathbf{a} \cdot \operatorname{grad} f \, \mathrm{d}V = \int_{\partial V} f\mathbf{a} \cdot \mathbf{n} \, \mathrm{d}S \qquad \text{using } \operatorname{div}\mathbf{a} = 0$$

$$\mathbf{a} \cdot \int_V \operatorname{grad} f \, \mathrm{d}V = \mathbf{a} \cdot \int_{\partial V} f\mathbf{n} \, \mathrm{d}S$$

Since $\mathbf{a}$ is arbitrary, we have verified our assertion. $\qquad\qquad\square$

Theorem 1.2 (Stokes' Theorem) *Let Σ be a surface bounded by the piecewise smooth, closed curve C, where $\mathbf{n}$ denotes the unit normal that forms a right-handed screw with the direction of circulation.[2] Further, let $\mathbf{u}$ be a vector field. Then*

$$\int_\Sigma (\operatorname{rot}\mathbf{u}) \cdot \mathbf{n} \, \mathrm{d}\Sigma = \oint_C \mathbf{u} \, \mathrm{d}\mathbf{x} \tag{1.26}$$

For the proof, we refer to any standard textbook on multidimensional analysis.

1.3   Fields and Potentials

A *gradient field* is a vector field $\mathbf{u}$ for which a scalar potential ϕ exists.

$$\mathbf{u} = \operatorname{grad}\phi$$

Both quantities $\mathbf{u}$ and ϕ can additionally depend on the time t.
A force field that possesses a potential is called *conservative*.

---

[2] If the thumb points along $\mathbf{n}$ and the index finger in the direction of the circulation, then the middle finger must point towards the surface.

Corollary 1.3 *The following statements are equivalent.*

(1) **u** *is a gradient field.*
(2) *The path integral of* **u** *does not depend on the chosen path $S(A, B)$, but only on its start and end point.*
(3) *The path integral of* **u** *vanishes around every closed curve:*

$$\oint \mathbf{u}\, d\mathbf{x} = 0$$

The proof can be found in standard texts on analysis, for example [2]. The path integral of a gradient field is calculated using the fundamental theorem of calculus for line integrals.

$$\int_{S(A,B)} \mathbf{u}\, d\mathbf{x} = \phi(B) - \phi(A)$$

Conservative forces play an important role in mechanics. An example is the gravitational force in the Earth's gravitational field. The work required to move a point mass corresponds to the increase in its potential energy, which only depends on the initial and final position. In contrast, frictional forces are not conservative, as the energy losses due to friction depend on the chosen path.

The existence of a potential depends not only on the vector field, but also on the topology of the region.

> A region is called *simply connected*, if every closed curve of the region can be continuously contracted to a point in the region.

Example 1.1 The exterior of the circle (Fig. 1.3 left) is multiply connected, as the enclosing curve bumps into the edge of the circle when contracted. In contrast, the exterior of the sphere (Fig. 1.3 right) is simply connected. The curve can be moved upwards so that it lies entirely in the exterior of the sphere. Now it can be contracted to a point.

> For a velocity field **u**, we denote the associated *vorticity* by
>
> $$\omega = \mathrm{rot}\,\mathbf{u}$$
>
> The vector field **u** is called *irrotational*, if
>
> $$\mathrm{rot}\,\mathbf{u} = 0 \tag{1.27}$$

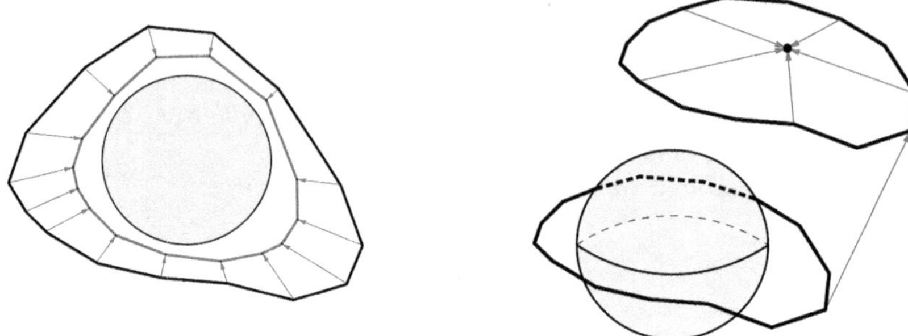

Fig. 1.3 (left) Exterior region in $\mathbb{R}^2$. (right) Exterior region in $\mathbb{R}^3$.

Important properties of the flow can be determined from the vorticity. If the latter are negligible or confined to boundary zones, we can apply mathematical methods that simplify the calculations.

Lemma 1.1 *Let* **u** *be a vector field on a region* Ω. *Then the following statements are true.*

(1) If **u** *is a gradient field, then it is irrotational.*

(2) If the region Ω *is simply connected and the vector field* **u** *is irrotational, then* **u** *is a gradient field.*

Proof

Ad. (1):   For $\mathbf{u} = \operatorname{grad} \phi$, it follows from (1.19) that

$$\operatorname{rot} \mathbf{u} = \operatorname{rot} \operatorname{grad} \phi = 0$$

Ad. (2):   Let $C \subset \Omega$ be a simple closed curve whose interior we denote by Σ. Since Ω is simply connected, it follows that $\Sigma \subset \Omega$ and we can apply Stokes' theorem (Theorem 1.2). Since $\operatorname{rot} \mathbf{u} = 0$ on Σ, the left-hand side of (1.26) vanishes and we find that $\int_C \mathbf{u}\,d\mathbf{x} = 0$, which verifies the gradient field according to Corollary 1.3.

$\square$

The requirement that Ω is simply connected turns out to be essential. A counterexample shows that an irrotational vector field does not possess a potential in any arbitrary region.

Problem 1.8 Prove the following statements for the vector field

$$\mathbf{u}(x, y) = \frac{1}{x^2 + y^2} \begin{pmatrix} -y \\ x \end{pmatrix}$$

in the multiply connected region $\mathbb{R}^2 \setminus \{0\}$.

(a) $\mathbf{u}$ is irrotational.
(b) When integrating around the unit circle, $\oint \mathbf{u} \, d\mathbf{x} \neq 0$.

Gradient fields will prove extremely useful in understanding aerodynamic lift in Chap. 4. In three-dimensional space, we model an irrotational airflow around a wing as a gradient field, which greatly simplifies the calculations. This is not readily possible in the analogous two-dimensional case because the flow-around region is no longer simply connected. Later we will see how, with an additional requirement on the two-dimensional velocity field, a gradient field is also obtained here.

The *Green's function* of the three-dimensional Laplace operator is defined by

$$G(\mathbf{x}, \mathbf{y}) = -\frac{1}{4\pi \, \|\mathbf{x} - \mathbf{y}\|} \qquad \text{for } \mathbf{x}, \mathbf{y} \in \mathbb{R}^3, \quad \mathbf{x} \neq \mathbf{y}$$

Furthermore, let

$$\mathbf{r} = \mathbf{y} - \mathbf{x}, \quad r = \|\mathbf{y} - \mathbf{x}\|, \quad \mathbf{e}_r = \frac{\mathbf{r}}{r}$$

Later, we will solve the Poisson equation $\Delta u = f$ using a potential approach and the Green's function. For this we need the following statements.

Problem 1.9 Let ∇_x and ∇_y denote the gradients with respect to $\mathbf{x}$ and $\mathbf{y}$, respectively. Prove the formulas

$$\nabla_y r = \frac{\mathbf{r}}{r} = \mathbf{e}_r \tag{1.28}$$

$$\nabla_x r = -\mathbf{e}_r \tag{1.29}$$

$$\nabla_x f(r) = -f'(r) \, \mathbf{e}_r \qquad \text{for scalar functions } f \tag{1.30}$$

Problem 1.10 In Green's function, we consider $\mathbf{x}$ as a parameter, i.e. G is only a function of $\mathbf{y}$. Show that the directional derivative is given by

$$\frac{dG(\mathbf{x}, \mathbf{y})}{d\mathbf{n}} = \frac{1}{4\pi} (\mathbf{n} \cdot \mathbf{e}_r) \, r^{-2} \tag{1.31}$$

References and Further Reading

1. Kelly, P.: *Solid Mechanics Lecture Notes, Part III: Foundations of Continuum Mechanics.* The University of Auckland, Auckland (2013), p. 119
   https://pkel015.connect.amazon.auckland.ac.nz/SolidMechanicsBooks/Part_III/index.html
   Accessed 3rd Sept 2024
2. Nikolsky, S.M.: *A Course of Mathematical Analysis.* Mir Publishers, Moscow (1981)

2.1 The Material Derivative

A neighbouring group of molecules that move at a certain average velocity at any given time is called a *particle*. In an idealised form, we reduce it to a point, the trajectory of which we follow.

A flow is determined by the location $\mathbf{x}(t)$ and the velocity $\mathbf{u}(\mathbf{x}(t), t)$ of its particles. In Cartesian coordinates, we denote the *trajectory* and *velocity* of a flowing particle by

$$\mathbf{x}(t) = \begin{pmatrix} x(t) \\ y(t) \\ z(t) \end{pmatrix}, \qquad \mathbf{u}(\mathbf{x}(t), t) = \begin{pmatrix} \dot{x}(t) \\ \dot{y}(t) \\ \dot{z}(t) \end{pmatrix} = \begin{pmatrix} u \\ v \\ w \end{pmatrix}$$

Now let $f(\mathbf{x}, t)$ be a scalar quantity, for example the density, which depends on *location and time*. Since physical quantities are bound to particles, it is useful to display their measured value and its change simultaneously with the moving particle. We can imagine a floating measuring device for this purpose. Mathematically, the

R. Spielmann, *Theoretical Fluid Mechanics*, https://doi.org/10.1007/978-3-662-72240-4_2

situation is captured by the function $f(\mathbf{x}(t), t)$ *along the flow*.[1] Its time derivative is

$$\frac{\mathrm{d}}{\mathrm{d}t} f(\mathbf{x}(t), t) = \frac{\partial f}{\partial t} + \frac{\partial f}{\partial x}\frac{\partial x}{\partial t} + \frac{\partial f}{\partial y}\frac{\partial y}{\partial t} + \frac{\partial f}{\partial z}\frac{\partial z}{\partial t}$$

$$= \frac{\partial f}{\partial t} + u\frac{\partial f}{\partial x} + v\frac{\partial f}{\partial y} + w\frac{\partial f}{\partial z} \tag{2.1}$$

The right side is called the *material derivative*. Instead of (2.1), one writes

$$\frac{Df}{Dt} = \frac{\partial f}{\partial t} + (\mathbf{u} \cdot \nabla)f \tag{2.2}$$

The last term in (2.2) is known as *directional derivative* of f in the direction of $\mathbf{u}$. It holds

$$(\mathbf{u} \cdot \nabla)f = \mathbf{u} \cdot \mathrm{grad}\, f$$

Thus, $\frac{D}{Dt}f = \frac{\mathrm{d}}{\mathrm{d}t}f(\mathbf{x}(t), t)$ denotes the temporal change of a physical quantity *following the flow*. It can be imagined as the change of the quantity when the measurement is coupled to a *moving particle*.

Examples of vectorial quantities include the vorticity $\boldsymbol{\omega}$ and the velocity $\mathbf{u}$. The *material derivative of a vector field* is formed component-wise according to the same principle:

$$\text{For}\quad \mathbf{v} = \begin{pmatrix} v_1 \\ v_2 \\ v_3 \end{pmatrix} \quad \text{we get}\quad \frac{D\mathbf{v}}{Dt} = \frac{\partial \mathbf{v}}{\partial t} + (\mathbf{u} \cdot \nabla)\mathbf{v} = \begin{pmatrix} \partial v_1/\partial t + (\mathbf{u} \cdot \nabla)v_1 \\ \partial v_2/\partial t + (\mathbf{u} \cdot \nabla)v_2 \\ \partial v_3/\partial t + (\mathbf{u} \cdot \nabla)v_3 \end{pmatrix}$$

The term $(\mathbf{u} \cdot \nabla)\mathbf{v}$ denotes our directional derivative of the vector field $\mathbf{v}$ in the direction of the velocity field $\mathbf{u}$ introduced in (1.12). It can be written as a matrix multiplication when we replace its components according to (2.1).

$$(\mathbf{u} \cdot \nabla)\mathbf{v} = \begin{pmatrix} \partial_x v_1 & \partial_y v_1 & \partial_z v_1 \\ \partial_x v_2 & \partial_y v_2 & \partial_z v_2 \\ \partial_x v_3 & \partial_y v_3 & \partial_z v_3 \end{pmatrix} \begin{pmatrix} u \\ v \\ u \end{pmatrix} \tag{2.3}$$

---

[1] Strictly speaking, the notation is incomplete, because the information about the starting value $\mathbf{x}(0) = \mathbf{a}$ should be added. Then the trajectory would be $\mathbf{x} = \mathbf{x}(\mathbf{a}, t)$ and the function of our measurement size $f = f(\mathbf{x}(\mathbf{a}, t), t)$. These designations are too cumbersome, so we only note the starting value $\mathbf{a}$ when it is absolutely necessary.

In particular, we obtain the acceleration as the material derivative of the velocity

$$\mathbf{a} = \frac{\partial \mathbf{u}}{\partial t} + (\mathbf{u} \cdot \nabla)\mathbf{u} \tag{2.4}$$

Problem 2.1 Prove the product rule $\frac{D}{Dt}(f\,g) = f\frac{Dg}{Dt} + g\frac{Df}{Dt}$.

2.2   Steady Flows and Streamlines

An *integral curve* of a vector field $\mathbf{u}$ is a curve that is tangential to the direction of the field at each of its points (see Fig. 2.1).

The *streamlines* at time t are the integral curves of the current velocity field $\mathbf{u}(\mathbf{x}, t)$.

A flow is called *steady* if $\frac{\partial \mathbf{u}}{\partial t} = 0$.

In a steady flow, the streamlines remain unchanged over time and correspond to the trajectories of the particles. The acceleration is time-independent at each point in space, because (2.4) implies $\mathbf{a} = (\mathbf{u} \cdot \nabla)\mathbf{u}$.

Fig. 2.1 Velocity field with associated integral curve.

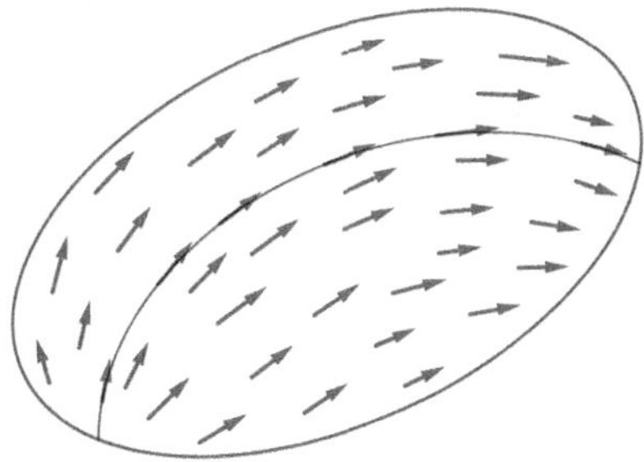

2.3   Mass Conservation and Continuity Equation

Let $\varrho = \varrho(\mathbf{x}, t)$ denote the density and $\mathbf{u} = \mathbf{u}(\mathbf{x}, t)$ the velocity of a fluid in a flow-through region V, which can be imagined as an anchored diving cage.

Theorem 2.1 (Continuity Equation) *In the entire flow region, it holds*

$$\frac{\partial \varrho}{\partial t} + \operatorname{div}(\varrho \mathbf{u}) = 0 \tag{2.5}$$

Proof Let us assume that at time t a fixed domain V of the flow region contains a fluid quantity of mass

$$m = \int_V \varrho(\mathbf{x}, t)\, dV$$

It changes only by inflow or outflow over the surface ∂V, i.e. it holds

$$\frac{d}{dt} \int_V \varrho(\mathbf{x}, t)\, dV = -\int_{\partial V} \varrho(\mathbf{x}, t)\, \mathbf{u}(\mathbf{x}, t) \cdot \mathbf{n}\, dS, \tag{2.6}$$

where $\mathbf{n}$ denotes the outward normal. For the left side of (2.6) we get

$$\frac{d}{dt} \int_V \varrho(\mathbf{x}, t)\, dV = \int_V \frac{\partial}{\partial t} \varrho(\mathbf{x}, t)\, dV,$$

since the temporal change is measured at a fixed location $\mathbf{x}$. The right integral in (2.6) can be transformed using Gauss's theorem (Theorem 1.1)

$$\int_{\partial V} \varrho(\mathbf{x}, t)\, \mathbf{u}(\mathbf{x}, t) \cdot \mathbf{n}\, dS = \int_V \operatorname{div}(\varrho \mathbf{u})\, dV$$

After inserting the simplifications into (2.6), we obtain

$$\int_V \frac{\partial \varrho}{\partial t}\, dV = -\int_V \operatorname{div}(\varrho \mathbf{u})\, dV$$

Since V is arbitrary, we have verified our assertion. $\square$

Corollary 2.1 *For a fluid with constant density ϱ, it holds*

$$\operatorname{div} \mathbf{u} = 0 \tag{2.7}$$

A vector field with property (2.7) is called *divergence-free* or *solenoidal*.

2.4   The Stream Function

We examine a two-dimensional, divergence-free flow, which can be unsteady. In the area $\bar{\Omega} \subset \mathbb{R}^2$ of the velocity field $\mathbf{u} = \begin{pmatrix} u(x, y, t) \\ v(x, y, t) \end{pmatrix}$ we consider the path $C(A, P)$ between the starting point A and the end point P. We choose $\mathbf{n}\,ds = \begin{pmatrix} dy \\ -dx \end{pmatrix}$ as the normally directed differential, see Problem 1.6. Then the curve integral

$$\psi_{C(A,P)} = \int_{C(A,P)} \mathbf{u} \cdot \mathbf{n}\,ds \tag{2.8}$$

describes the *flow rate* at time t, i.e. the quantity flowing through our curve per unit of time (see Fig. 2.2).

Problem 2.2 Show that the value $\psi_{C(A,P)}$ in a divergence-free flow depends only on the starting and end point, but not on the chosen path.

This justifies the following definition.

Let $A \in \bar{\Omega}$ be a fixed point in the flow area. The assignment

$$\psi(x, y) = \int_{C(A,P)} \mathbf{u} \cdot \mathbf{n}\,ds \quad \text{where } P \in \bar{\Omega} \text{ has the coordinates } (x, y)$$

is called the *stream function*. In unsteady flows, it depends on t.

Stream functions for different starting points A differ by an added constant. A stationary velocity field $\mathbf{u} = \begin{pmatrix} u \\ v \end{pmatrix}$ can be calculated from its (time-independent) stream function $\psi(x, y)$ as follows.

Fig. 2.2 Flow through an arc $C(A, P)$.

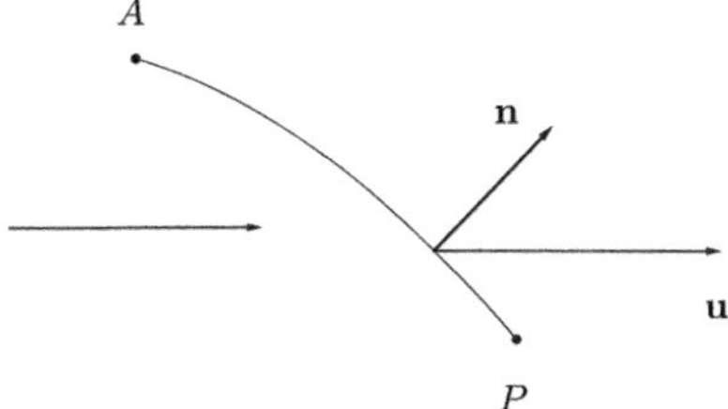

Lemma 2.1 *In the stationary case, the partial derivatives of the stream function are*

$$u = \frac{\partial \psi}{\partial y}, \qquad v = -\frac{\partial \psi}{\partial x} \qquad (2.9)$$

Proof We choose a path $C(A, P)$ with a parametrisation

$$\mathbf{x}(\tau) = \begin{pmatrix} x(\tau) \\ y(\tau) \end{pmatrix}, \qquad \tau \in [0, \eta]$$

Then $\mathbf{u} \cdot \mathbf{n}\, ds = u\, dy - v\, dx$ and thus

$$\psi(x(\eta), y(\eta)) = \int_0^\eta \left(u \frac{dy}{d\tau} - v \frac{dx}{d\tau} \right) d\tau$$

$$\frac{d\psi}{d\eta} = u \frac{dy}{d\eta} - v \frac{dx}{d\eta}$$

$$\frac{\partial \psi}{\partial x} \frac{dx}{d\eta} + \frac{\partial \psi}{\partial y} \frac{dy}{d\eta} = u \frac{dy}{d\eta} - v \frac{dx}{d\eta}$$

By comparing coefficients we obtain the desired relations. $\square$

Problem 2.3 Let $\mathbf{u}$ be a stationary, divergence-free velocity field on a domain $\Omega \subset \mathbb{R}^2$. Prove that the stream function ψ remains constant on each streamline.

In lift calculations for airfoil profiles $D \subset \mathbb{R}^2$, we consider exterior regions, whose boundary ∂D forms a connected curve. Since the fluid does not penetrate the boundary, it follows from (2.8) that

$$\psi = \text{const} \qquad \text{on } \partial \Omega \qquad (2.10)$$

In general, this level set consists of several streamlines, which are separated by stagnation points.

Further, from (2.9) we get the relations $\frac{\partial^2 \psi}{\partial x^2} = -\frac{\partial v}{\partial x}$, $\frac{\partial^2 \psi}{\partial y^2} = \frac{\partial u}{\partial y}$. After addition and taking into account (1.21), we obtain

$$\Delta \psi = -\text{rot}\, \mathbf{u} \qquad \text{in } \Omega \qquad (2.11)$$

Thus, a stream function for a stationary, divergence-free velocity field $\mathbf{u}$ can be calculated as a solution of the Poisson equation (2.10) and (2.11).

Corollary 2.2 *For a given stationary, divergence-free velocity field $\mathbf{u}$, a stream function ψ can be determined from equations (2.9) and (2.10).*

2.5   The Transport Theorem

We want to examine the temporal change of a physical quantity in a "co-moving" spatial region.

> By a *flowing volume* V_t we mean a *region*, i.e. an open, non-empty and connected subset in three-dimensional space that is filled with flowing particles at time t.

We can imagine it as the current location of a floating liquid portion that has been marked with a dye.

Let **a** be the starting point of a particle at time zero. After time t, it reaches the location $\mathbf{x} = \mathbf{x}(\mathbf{a}, t)$. The assignment of all initial locations to their locations reached after a fixed time t

$$M_t : \mathbf{a} \mapsto \mathbf{x}(\mathbf{a}, t)$$

is called the *Lagrange map* (Fig. 2.3). Its *Jacobian matrix*

$$DM_t = \begin{pmatrix} \frac{\partial x_1}{\partial a_1} & \frac{\partial x_1}{\partial a_2} & \frac{\partial x_1}{\partial a_3} \\ \frac{\partial x_2}{\partial a_1} & \frac{\partial x_2}{\partial a_2} & \frac{\partial x_2}{\partial a_3} \\ \frac{\partial x_3}{\partial a_1} & \frac{\partial x_3}{\partial a_2} & \frac{\partial x_3}{\partial a_3} \end{pmatrix} = \left(\frac{\partial x_i}{\partial a_j} \Big|_t \right) \tag{2.12}$$

and the associated *Jacobian determinant*

$$\det DM_t = \epsilon_{ijk} \frac{\partial x_1}{\partial a_i} \frac{\partial x_2}{\partial a_j} \frac{\partial x_3}{\partial a_k} \tag{2.13}$$

allow us to reduce integrals over "flowing" subregions V_t to integrals over the initial region V_0. The latter can be derived as usual after time. First, we examine how the Lagrange map changes the volumes of flow areas.

Using the box product (1.7) of the vectors

$$\mathbf{da}^{(1)} = \begin{pmatrix} da_1 \\ 0 \\ 0 \end{pmatrix}, \qquad \mathbf{da}^{(2)} = \begin{pmatrix} 0 \\ da_2 \\ 0 \end{pmatrix}, \qquad \mathbf{da}^{(3)} = \begin{pmatrix} 0 \\ 0 \\ da_3 \end{pmatrix},$$

we define an infinitesimal volume element at the initial time

$$dV_0 = \mathbf{da}^{(1)} \cdot (\mathbf{da}^{(2)} \times \mathbf{da}^{(3)}) = da_1 \, da_2 \, da_3 \tag{2.14}$$

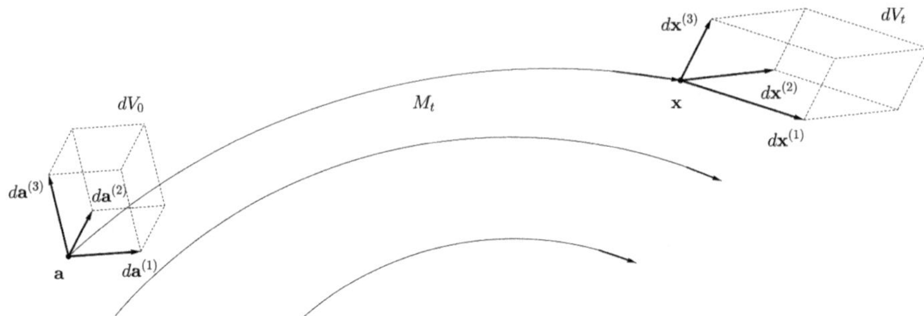

Fig. 2.3 Lagrange map M_t.

The Jacobian matrix (2.12) transforms the vectors $\mathbf{da}^{(k)}$ into their images

$$\mathbf{dx}^{(k)} = DM_t\, \mathbf{da}^{(k)}$$

In particular, we have

$$
\mathbf{dx}^{(1)} = \begin{pmatrix} \frac{\partial x_1}{\partial a_1}\, da_1 \\[2mm] \frac{\partial x_2}{\partial a_1}\, da_1 \\[2mm] \frac{\partial x_3}{\partial a_1}\, da_1 \end{pmatrix}, \qquad
\mathbf{dx}^{(2)} = \begin{pmatrix} \frac{\partial x_1}{\partial a_2}\, da_2 \\[2mm] \frac{\partial x_2}{\partial a_2}\, da_2 \\[2mm] \frac{\partial x_3}{\partial a_2}\, da_2 \end{pmatrix}, \qquad
\mathbf{dx}^{(3)} = \begin{pmatrix} \frac{\partial x_1}{\partial a_3}\, da_3 \\[2mm] \frac{\partial x_2}{\partial a_3}\, da_3 \\[2mm] \frac{\partial x_3}{\partial a_3}\, da_3 \end{pmatrix}
$$

Now we can calculate the desired volume dV_t. Using the box product, taking into account (2.12) and (2.14), we find

$$
dV_t = \mathbf{dx}^{(1)} \cdot (\mathbf{dx}^{(2)} \times \mathbf{dx}^{(3)}) = \epsilon_{ijk}\, \frac{\partial x_1}{\partial a_i}da_i\, \frac{\partial x_2}{\partial a_j}da_j\, \frac{\partial x_3}{\partial a_k}da_k = (\det DM_t)\, dV_0
$$

$$(2.15)$$

Now we want to examine the temporal change of the Jacobian determinant. Considering $\frac{d}{dt}\frac{\partial x_n}{\partial a_m} = \frac{\partial u_n}{\partial a_m}$, we get from (2.13)

$$
\frac{d}{dt}\det DM_t = \frac{d}{dt}\left(\epsilon_{ijk}\,\frac{\partial x_1}{\partial a_i}\frac{\partial x_2}{\partial a_j}\frac{\partial x_3}{\partial a_k}\right)
$$

$$
= \epsilon_{ijk}\,\frac{\partial u_1}{\partial a_i}\frac{\partial x_2}{\partial a_j}\frac{\partial x_3}{\partial a_k} + \epsilon_{ijk}\,\frac{\partial x_1}{\partial a_i}\frac{\partial u_2}{\partial a_j}\frac{\partial x_3}{\partial a_k} + \epsilon_{ijk}\,\frac{\partial x_1}{\partial a_i}\frac{\partial x_2}{\partial a_j}\frac{\partial u_3}{\partial a_k} \quad (2.16)
$$

In expanded form, the first term in the last line yields

$$\epsilon_{ijk}\frac{\partial u_1}{\partial a_i}\frac{\partial x_2}{\partial a_j}\frac{\partial x_3}{\partial a_k} = \det\begin{bmatrix} \frac{\partial u_1}{\partial a_1} & \frac{\partial u_1}{\partial a_2} & \frac{\partial u_1}{\partial a_3} \\ \frac{\partial x_2}{\partial a_1} & \frac{\partial x_2}{\partial a_2} & \frac{\partial x_2}{\partial a_3} \\ \frac{\partial x_3}{\partial a_1} & \frac{\partial x_3}{\partial a_2} & \frac{\partial x_3}{\partial a_3} \end{bmatrix} \tag{2.17}$$

Furthermore, using summation convention in the last term respectively

$$\frac{\partial u_1}{\partial a_1} = \frac{\partial u_1}{\partial x_1}\frac{\partial x_1}{\partial a_1} + \frac{\partial u_1}{\partial x_2}\frac{\partial x_2}{\partial a_1} + \frac{\partial u_1}{\partial x_3}\frac{\partial x_3}{\partial a_1} = \frac{\partial u_1}{\partial x_i}\frac{\partial x_i}{\partial a_1}$$

$$\frac{\partial u_1}{\partial a_2} = \frac{\partial u_1}{\partial x_1}\frac{\partial x_1}{\partial a_2} + \frac{\partial u_1}{\partial x_2}\frac{\partial x_2}{\partial a_2} + \frac{\partial u_1}{\partial x_3}\frac{\partial x_3}{\partial a_2} = \frac{\partial u_1}{\partial x_i}\frac{\partial x_i}{\partial a_2}$$

$$\frac{\partial u_1}{\partial a_3} = \frac{\partial u_1}{\partial x_1}\frac{\partial x_1}{\partial a_3} + \frac{\partial u_1}{\partial x_2}\frac{\partial x_2}{\partial a_3} + \frac{\partial u_1}{\partial x_3}\frac{\partial x_3}{\partial a_3} = \frac{\partial u_1}{\partial x_i}\frac{\partial x_i}{\partial a_3}$$

If we replace these relations in the top row of (2.17), we get

$$\epsilon_{ijk}\frac{\partial u_1}{\partial a_i}\frac{\partial x_2}{\partial a_j}\frac{\partial x_3}{\partial a_k} = \det\begin{bmatrix} \frac{\partial u_1}{\partial x_i}\frac{\partial x_i}{\partial a_1} & \frac{\partial u_1}{\partial x_i}\frac{\partial x_i}{\partial a_2} & \frac{\partial u_1}{\partial x_i}\frac{\partial x_i}{\partial a_3} \\ \frac{\partial x_2}{\partial a_1} & \frac{\partial x_2}{\partial a_2} & \frac{\partial x_2}{\partial a_3} \\ \frac{\partial x_3}{\partial a_1} & \frac{\partial x_3}{\partial a_2} & \frac{\partial x_3}{\partial a_3} \end{bmatrix} \tag{2.18}$$

According to Problem 1.2, the value of a determinant remains unchanged if we subtract the second row multiplied by $\frac{\partial u_1}{\partial x_2}$ and the third row multiplied by $\frac{\partial u_1}{\partial x_3}$ from the first row. Subsequently, by Problem 1.3, we can place the factor $\frac{\partial u_1}{\partial x_1}$ from the first row in front of the determinant. This simplifies (2.18) to

$$\epsilon_{ijk}\frac{\partial u_1}{\partial a_i}\frac{\partial x_2}{\partial a_j}\frac{\partial x_3}{\partial a_k} = \det\begin{bmatrix} \frac{\partial u_1}{\partial x_1}\frac{\partial x_1}{\partial a_1} & \frac{\partial u_1}{\partial x_1}\frac{\partial x_1}{\partial a_2} & \frac{\partial u_1}{\partial x_1}\frac{\partial x_1}{\partial a_3} \\ \frac{\partial x_2}{\partial a_1} & \frac{\partial x_2}{\partial a_2} & \frac{\partial x_2}{\partial a_3} \\ \frac{\partial x_3}{\partial a_1} & \frac{\partial x_3}{\partial a_2} & \frac{\partial x_3}{\partial a_3} \end{bmatrix} = \frac{\partial u_1}{\partial x_1}\det\begin{bmatrix} \frac{\partial x_1}{\partial a_1} & \frac{\partial x_1}{\partial a_2} & \frac{\partial x_1}{\partial a_3} \\ \frac{\partial x_2}{\partial a_1} & \frac{\partial x_2}{\partial a_2} & \frac{\partial x_2}{\partial a_3} \\ \frac{\partial x_3}{\partial a_1} & \frac{\partial x_3}{\partial a_2} & \frac{\partial x_3}{\partial a_3} \end{bmatrix}$$

$$= \frac{\partial u_1}{\partial x_1}\det DM_t$$

Similarly, we obtain for the remaining terms of (2.16)

$$\epsilon_{ijk}\frac{\partial x_1}{\partial a_i}\frac{\partial u_2}{\partial a_j}\frac{\partial x_3}{\partial a_k} = \frac{\partial u_2}{\partial x_2}\det DM_t, \qquad \epsilon_{ijk}\frac{\partial x_1}{\partial a_i}\frac{\partial x_2}{\partial a_j}\frac{\partial u_3}{\partial a_k} = \frac{\partial u_3}{\partial x_3}\det DM_t$$

In summary, it follows from (2.16) that

$$\frac{\mathrm{d}}{\mathrm{d}t}\det DM_t = \Big(\frac{\partial u_1}{\partial x_1} + \frac{\partial u_2}{\partial x_2} + \frac{\partial u_3}{\partial x_3}\Big)\det DM_t = (\operatorname{div}\mathbf{u})\det DM_t \qquad (2.19)$$

Theorem 2.2 (Reynolds Transport Theorem) *Let V_t be a flowing volume and $f = f(\mathbf{x}(t), t)$ a scalar function.* [2] *Then it holds*

$$\frac{\mathrm{d}}{\mathrm{d}t}\int_{V_t} f\,\mathrm{d}V_t = \int_{V_t}\left[\frac{Df}{Dt} + f\operatorname{div}\mathbf{u}\right]\mathrm{d}V_t$$

Proof Using the Lagrange mapping, we transform the integral back. By (2.15) this yields

$$\frac{\mathrm{d}}{\mathrm{d}t}\int_{V_t} f\,\mathrm{d}V_t = \frac{\mathrm{d}}{\mathrm{d}t}\int_{V_0} f(\mathbf{x}(t), t)\det DM_t\,\mathrm{d}V_0$$

$$= \int_{V_0}\frac{\mathrm{d}}{\mathrm{d}t}[f(\mathbf{x}(t), t)\det DM_t]\,\mathrm{d}V_0 \qquad (2.20)$$

Applying the product rule, taking into account the material derivative (2.1) and (2.19), we find that

$$\frac{\mathrm{d}}{\mathrm{d}t}[f(\mathbf{x}(t), t)\det DM_t] = \frac{Df}{Dt}\det DM_t + f(\mathbf{x}(t), t)(\operatorname{div}\mathbf{u})\det DM_t$$

$$= \left[\frac{Df}{Dt} + f\operatorname{div}\mathbf{u}\right]\det DM_t$$

Substituting into (2.20) we get

$$\frac{\mathrm{d}}{\mathrm{d}t}\int_{V_t} f\,\mathrm{d}V_t = \int_{V_0}\left[\frac{Df}{Dt} + f\operatorname{div}\mathbf{u}\right]\det DM_t\,\mathrm{d}V_0$$

$$= \int_{V_t}\left[\frac{Df}{Dt} + f\operatorname{div}\mathbf{u}\right]\mathrm{d}V_t \qquad \text{by (2.15)}$$

$$\square$$

---

[2] The function f is considered "along the flow". Its values can be imagined as a series of measurements of a corresponding physical quantity, with the measuring device floating along with the particle stream.

A fluid is *incompressible* when its volume remains constant despite the action of a force or a change in pressure.

Corollary 2.3 *A fluid is incompressible if and only if* $\operatorname{div}\mathbf{u} = 0$.

Proof As usual, $V_t \subset \mathbb{R}^3$ denotes the set of points and $|V_t|$ the measure of its volume. Incompressibility means for every flowing volume V_t that

$$\frac{\mathrm{d}|V_t|}{\mathrm{d}t} = \frac{\mathrm{d}}{\mathrm{d}t}\int_{V_t}\mathrm{d}V_t = 0$$

Using the transport theorem, we find that $\frac{\mathrm{d}}{\mathrm{d}t}\int_{V_t}\mathrm{d}V_t = \int_{V_t}\operatorname{div}\mathbf{u}\,\mathrm{d}V_t$. It follows that $\int_{V_t}\operatorname{div}\mathbf{u}\,\mathrm{d}V_t = 0$, thus $\operatorname{div}\mathbf{u} = 0$. $\square$

In most cases, liquids can be considered incompressible. The same applies to gases, as long as the flow velocity is small compared to the speed of sound[3] and the temperature remains constant.

Example 2.1 Owing to incompressibility, the same mass flows through each cross-section S of a pipe per unit of time, i.e. we get

$$\varrho u S = \text{const}$$

For constant fluid density, it follows that $u_2 = u_1 S_1 / S_2$, which is the reason why the speed increases at constrictions (see Fig. 2.4).

This situation can even be observed in weather phenomena (see Fig. 2.5).

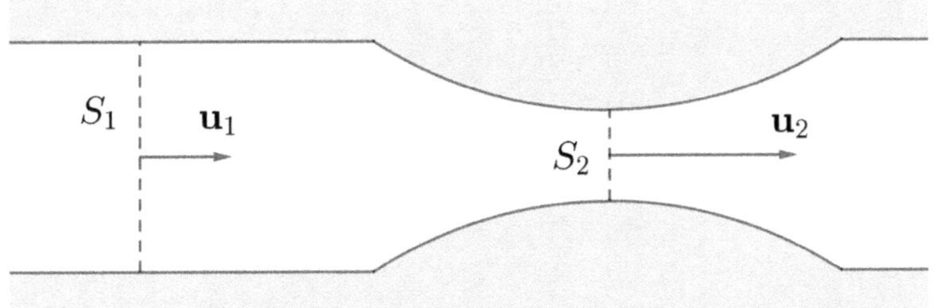

Fig. 2.4 Flow through a constricting pipe.

---

[3] The speed of sound in air is about 350 m/s.

Fig. 2.5 The Mistral is a wind that blows from the south of France into the Gulf of Lion in the northern Mediterranean. In the Rhone Valley, it is narrowed between the mountains of the Massif Central and the Alps, so that the wind speed increases there and reaches peak values of around 135 km/h. The cold, dry wind removes moisture from the soil and shapes the vegetation of the entire region. Image: NASA.

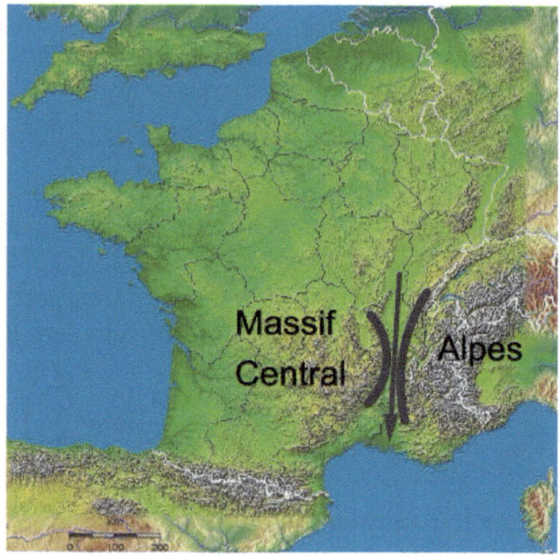

We recall that a fluid with constant density is incompressible by Corollary 2.1. However, incompressibility does not automatically mean that the density remains constant. Seawater can be considered incompressible, although its density varies due to different salt content.

References and Further Reading

1. Chorin A.J., Marsden J.E.: *A Mathematical Introduction to Fluid Mechanics.* Springer, Berlin (1992)
2. Milne-Thomson L.M.: *Theoretical Hydrodynamics.* Dover Publications, New York (1968)

Dynamics of Flows

3

Flows in liquids and gases are modeled by the Navier-Stokes equations. Their derivation is carried out in several steps, with the most important partial results being summarised in auxiliary theorems. Our starting point is Newton's second law

$$\frac{\mathrm{d}}{\mathrm{d}t} m\mathbf{v} = \mathbf{F} \tag{3.1}$$

Since it is applied to point masses or bodies, its fluid mechanical variant must find a corresponding application to flowing particles or volumes V_t.

3.1 Internal and External Forces

The force on a flowing volume V_t is composed of two components. *Internal forces* $\mathbf{F}_i$ arise from the interaction of particles with each other, in particular due to pressure and frictional resistance of neighbouring particles. These are contact forces that act via surfaces. The internal force acting on V_t can be written as a surface integral with the matrix of the *stress tensor* $\mathsf{T} = (T_{ij})$.

$$\mathbf{F}_i = \int_{\partial V_t} \mathsf{T} \cdot \mathbf{n} \, \mathrm{d}S, \quad \mathbf{n} \text{ outward pointing surface normal to } \partial V \tag{3.2}$$

In elasticity theory and fluid mechanics, *stress* is defined as force per area. The integral (3.2) sums over all stresses $\mathsf{T} \cdot \mathbf{n}$ on the surface ∂V_t. It summarises the force acting on V_t caused by the particles located outside V_t. Within V_t, the interaction forces cancel each other out, so they do not contribute to the movement of V_t.

External forces $\mathbf{F}_a$ arise from effects mediated by external fields. Their causes lie outside the considered system. Examples are gravitational forces or electromagnetic forces in liquid metals. They act on each particle in V_t and can be described as a

R. Spielmann, *Theoretical Fluid Mechanics*, https://doi.org/10.1007/978-3-662-72240-4_3

volume integral using a force density $\mathbf{f}$.

$$\mathbf{F}_a = \int_{V_t} \mathbf{f}\, dV \tag{3.3}$$

The forces lead to a change in momentum, which is formulated as follows.

Lemma 3.1 *In a flow, Newton's second law takes the form*

$$\varrho \frac{D\mathbf{u}}{Dt} = \mathbf{f} + \operatorname{div} \mathsf{T} \tag{3.4}$$

Proof The left-hand side of (3.1) describes the *change in momentum*. For any flowing volume V_t it is calculated as a volume integral

$$\frac{d}{dt} \int_{V_t} \varrho \mathbf{u}\, dV$$

The force $\mathbf{F}$ on the right-hand side of Newton's equation (3.1) is composed of internal and external forces. Thus, for V_t, Newton's equation (3.1) takes the following form.

$$\frac{d}{dt} \int_{V_t} \varrho \mathbf{u}\, dV = \int_{V_t} \mathbf{f}\, dV + \int_{\partial V_t} \mathsf{T} \cdot \mathbf{n}\, dS \tag{3.5}$$

On applying the transport theorem (Theorem 2.2) to the left-hand side and Gauss's theorem (Theorem 1.1) to the surface integral, we obtain

$$\int_{V_t} \left[\frac{D}{Dt}(\varrho \mathbf{u}) + (\varrho \mathbf{u}) \operatorname{div} \mathbf{u} \right] dV = \int_{V_t} \mathbf{f}\, dV + \int_{V_t} \operatorname{div} \mathsf{T}\, dV$$

Since V_t can be chosen arbitrarily, it follows that

$$\frac{D}{Dt}(\varrho \mathbf{u}) + (\varrho \mathbf{u}) \operatorname{div} \mathbf{u} = \mathbf{f} + \operatorname{div} \mathsf{T} \tag{3.6}$$

Using the product rule (Exercise 2.1) and the material derivative (2.2), we get

$$\frac{D}{Dt}(\varrho \mathbf{u}) = \varrho \frac{D\mathbf{u}}{Dt} + \mathbf{u} \frac{D\varrho}{Dt} = \varrho \frac{D\mathbf{u}}{Dt} + \mathbf{u} \left[\frac{\partial \varrho}{\partial t} + \mathbf{u} \cdot \nabla \varrho \right]$$

We simplify the left-hand side of (3.6)

$$\frac{D}{Dt}(\varrho\mathbf{u}) + (\varrho\mathbf{u})\operatorname{div}\mathbf{u} = \varrho\frac{D\mathbf{u}}{Dt} + \mathbf{u}\left[\frac{\partial\varrho}{\partial t} + \mathbf{u}\cdot\nabla\varrho + \varrho\operatorname{div}\mathbf{u}\right]$$

$$= \varrho\frac{D\mathbf{u}}{Dt} + \mathbf{u}\left[\frac{\partial\varrho}{\partial t} + \operatorname{div}(\varrho\mathbf{u})\right] \qquad \text{by (1.15)}$$

$$= \varrho\frac{D\mathbf{u}}{Dt} \qquad \text{using the continuity equation (2.5)}$$

By substitution into (3.6) we obtain the statement. $\qquad\square$

3.2 Pressure and Shear Stress

In (3.2), $\mathsf{T}\cdot\mathbf{n}$ denotes the stress (=force per area) on a surface element with unit normal $\mathbf{n}$. To simplify the stress tensor $\mathsf{T} = (T_{ij})$, we use the approach

$$\mathsf{T} = -p\mathsf{I} + \mathsf{S} \tag{3.7}$$

$$\text{or} \quad T_{ij} = -p\,\delta_{ij} + \sigma_{ij}$$

Here p denotes the pressure, $\mathsf{I} = (\delta_{ij})$ the unit tensor and $\mathsf{S} = (\sigma_{ij})$ the *tensor of shear stress*, also called *viscous stress tensor*. For the latter, the diagonal elements vanish, i.e.

$$\sigma_{ii} = 0 \tag{3.8}$$

Both the scalar p and the tensor components T_{ij}, σ_{ij} are functions of $\mathbf{x}$ and t, whereby the pressure is always positive. To interpret the pressure component, we consider (3.2) with vanishing shear stress, i.e.

$$\mathbf{F}_i = -\int_{\partial V_t} p\,\mathsf{I}\cdot\mathbf{n}\,\mathrm{d}S = -\int_{\partial V_t} p\mathbf{n}\,\mathrm{d}S$$

We have obtained a formula for the force acting on the body V_t, which is caused by the pressure on its surface. The pressure on a surface element acts opposite to the external surface normal $\mathbf{n}$, thus *perpendicular to the surface* (Fig. 3.1 left).

A fluid whose internal forces are only caused by pressure is called an *ideal fluid* and is inviscid.

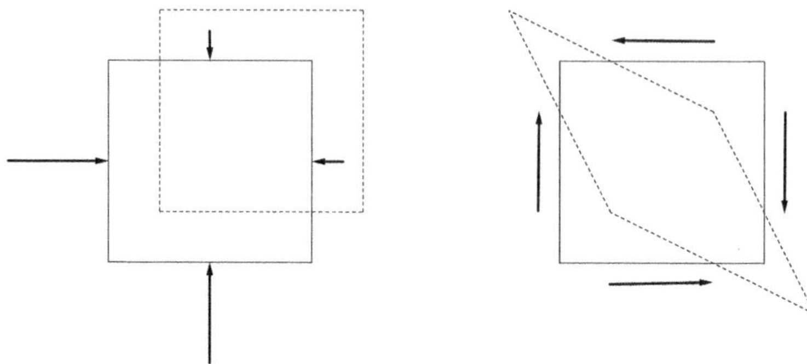

Fig. 3.1 (left) Pressure forces act perpendicular to the surface. (right) Shear forces act parallel to the surface. Dashed lines show the altered body.

Forces due to shear stresses act *parallel to the surface* and cause distortions (Fig. 3.1 right). This creates friction.

Problem 3.1 Prove that the shear force to the plane with the normal $\mathbf{e}_i$ acts parallel to this plane.

Observations show that the particles of a flow are not automatically forced into rapid rotational movements. From this, we can derive a symmetry.

Lemma 3.2 *The stress tensor* T *is symmetric, i.e.*

$$T_{ij} = T_{ji} \tag{3.9}$$

Proof The force generates a torque, which changes the angular momentum of the particle. In analogy with (3.5),

$$\frac{\mathrm{d}}{\mathrm{d}t} \int_{V_t} \varrho (\mathbf{x} \times \mathbf{u}) \, \mathrm{d}V = \int_{V_t} (\mathbf{x} \times \mathbf{f}) \, \mathrm{d}V + \int_{\partial V_t} (\mathbf{x} \times [\mathsf{T} \cdot \mathbf{n}]) \, \mathrm{d}S \tag{3.10}$$

First, we examine the torque

$$\mathbf{M} = \int_{V_t} (\mathbf{x} \times \mathbf{f}) \, \mathrm{d}V + \int_{\partial V_t} (\mathbf{x} \times [\mathsf{T} \cdot \mathbf{n}]) \, \mathrm{d}S \tag{3.11}$$

Using the notation $\mathsf{R} = (R_{kq}) = (\epsilon_{ijk} \, x_i \, T_{jq})$, we obtain from (1.3)

$$\mathbf{x} \times [\mathsf{T} \cdot \mathbf{n}] = \epsilon_{ijk} \, x_i (T_{jq} n_q) = \mathsf{R} \cdot \mathbf{n}$$

Applying Gauss's theorem (Theorem 1.1),

$$\int_{\partial V_t} (\mathbf{x} \times [\mathsf{T} \cdot \mathbf{n}]) \, \mathrm{d}S = \int_{\partial V_t} \mathsf{R} \cdot \mathbf{n} \, \mathrm{d}S = \int_{V_t} \mathrm{div} \, \mathsf{R} \, \mathrm{d}V$$

and thus

$$\mathbf{M} = \int_{V_t} (\mathbf{x} \times \mathbf{f} + \operatorname{div} \mathsf{R}) \, \mathrm{d}V$$

The integrand has the coordinates

$$
\begin{aligned}
\mathbf{x} \times \mathbf{f} + \operatorname{div} \mathsf{R} &= \epsilon_{ijk}\, x_i\, f_j + \partial_q (\epsilon_{ijk}\, x_i\, T_{jq}) \\
&= \epsilon_{ijk}\, x_i\, f_j + \epsilon_{ijk}\, x_i\, \partial_q T_{jq} + \epsilon_{ijk}\, T_{jq}\, \partial_q x_i \\
&= \epsilon_{ijk}\, x_i\, f_j + \epsilon_{ijk}\, x_i\, \partial_q T_{jq} + \epsilon_{ijk}\, T_{jq}\, \delta_{qi} \\
&= \epsilon_{ijk}\, x_i\, f_j + \epsilon_{ijk}\, x_i\, \partial_q T_{jq} + \epsilon_{ijk}\, T_{ji}
\end{aligned}
$$

and we obtain the torque

$$\mathbf{M} = \int_{V_t} \epsilon_{ijk}\, x_i\, f_j \, \mathrm{d}V + \int_{V_t} \epsilon_{ijk}\, x_i\, \partial_q T_{jq} \, \mathrm{d}V + \int_{V_t} \epsilon_{ijk}\, T_{ji} \, \mathrm{d}V \tag{3.12}$$

By $|V_t|$ we denote the measure of the volume $V_t \subset \mathbb{R}^3$. To investigate small particles, we reduce V_t while maintaining its proportions. For sufficiently small volumes, we can approximately consider the density as well as the components of the force and the stress tensor as constant, whereas for spatial coordinates $x_i = O(|V|^{1/3})$. This allows us to estimate the orders of the terms in (3.12) for $|V_t| \to 0$ (see Sect. A.1). Since there are no spatial coordinates under the last integral, it holds

$$\int_{V_t} \epsilon_{ijk}\, T_{ji} \, \mathrm{d}V = O(|V_t|) \tag{3.13}$$

The remaining integrals in (3.12) contain x_i and are therefore of the order $O(|V_t|^{4/3})$. Now let's consider the left side of (3.10). It describes the change in angular momentum

$$\frac{\mathrm{d}\mathbf{L}}{\mathrm{d}t} = \frac{\mathrm{d}}{\mathrm{d}t} \int_{V_t} \varrho\,(\mathbf{x} \times \mathbf{u}) \, \mathrm{d}V$$

The integrand contains the position vector $\mathbf{x} = O(|V_t|^{1/3})$. Then

$$\frac{\mathrm{d}\mathbf{L}}{\mathrm{d}t} = O(|V_t|^{4/3}) \tag{3.14}$$

If the change in angular momentum (3.14) for $|V_t| \to 0$ tends more strongly to zero than the generating torque (3.13), then the angular acceleration changes with the order $O(|V_t|/|V_t|^{4/3}) = O(|V_t|^{-1/3}) \to \infty$. Small particles would have to rotate faster and faster, which contradicts observation. Consequently, the integral (3.13) must not contribute, i.e. we get $\epsilon_{ijk} T_{ji} = 0$. This means

$$T_{12} - T_{21} = 0, \quad T_{32} - T_{23} = 0, \quad T_{13} - T_{31} = 0$$

$\square$

By (3.7), the shear stress tensor S is also symmetric. To determine its structure, we first consider the matrix $\nabla\mathbf{u}$ of the first derivatives of the velocity. Through the decomposition [1]

$$\nabla\mathbf{u} = \frac{1}{2}(\nabla\mathbf{u} + (\nabla\mathbf{u})^T) + \frac{1}{2}(\nabla\mathbf{u} - (\nabla\mathbf{u})^T)$$

we obtain $\nabla\mathbf{u} = \mathsf{E} + \Omega$ as the sum of a symmetric part

$$\mathsf{E} = \frac{1}{2}(\nabla\mathbf{u} + (\nabla\mathbf{u})^T) \tag{3.15}$$

$$\text{or} \quad e_{ij} = \frac{1}{2}\left(\frac{\partial u_i}{\partial x_j} + \frac{\partial u_j}{\partial x_i}\right)$$

and an asymmetric part

$$\Omega = \frac{1}{2}(\nabla\mathbf{u} - (\nabla\mathbf{u})^T)$$

The symmetric tensor E is known as the *deformation tensor*.

3.3 Shear Stress and Velocity Gradient

Based on further observations, we can refine the approach. In constant velocity fields, no shear forces occur. Moreover, increasing the flow velocity by an additive constant cannot lead to deformation of particles. Thus, S does not depend directly on $\mathbf{u}$, but is essentially determined by its first derivatives $\partial u_i/\partial x_j$.

Assumption 1 The shear stress tensor S depends linearly on $\nabla\mathbf{u}$. Higher derivatives of the velocity do not enter into S.

The neglect of higher derivatives of $\mathbf{u}$ as well as the assumption of linearity are convenient simplifications, but they cannot always be made (see remark at the end

---

[1] By $\mathbf{A}^T = (a_{ji})$ we denote the transpose of the matrix $\mathbf{A} = (a_{ij})$.

of the section). If we denote a corresponding fluid as a *Newtonian fluid*, we obtain
the representation

$$S = A(\nabla \mathbf{u}) \tag{3.16}$$

$$\text{or} \quad \sigma_{ij} = A_{ijkm} \frac{\partial u_k}{\partial x_m},$$

where $A = (A_{ijkm})$ is a fourth-order tensor.

> **Assumption 2**   The fluid is *homogeneous*, i.e. the tensor S does not explicitly
> depend on the spatial coordinates x_i.

This statement is based on the fact that only the flow at the point in space, but
not the point itself, determines the shear forces. Thus, our tensor A has constant
coefficients.

> **Assumption 3**   The fluid is *isotropic*, i.e. there is no preferred spatial
> direction.

If a velocity field were rotated by a fixed angle, all quantities calculated from $\nabla \mathbf{u}$
would be rotated by the same angle. This includes the shear force determined by
(3.16). Therefore, the coordinates of A are not changed by rotation. Such tensors are
called *isotropic*. For examples, see Sect. A.2.

Lemma 3.3 *An isotropic fourth-order tensor can be written in the form*

$$A_{ijkm} = \alpha \delta_{ij} \delta_{km} + \mu \delta_{ik} \delta_{jm} + \gamma \delta_{im} \delta_{jk}$$

where α, μ and γ are constants.

A proof can be found in [5].

For the shear stress $S = (\sigma_{ij})$, it follows from (3.16) that

$$\sigma_{ij} = (\alpha \delta_{ij} \delta_{km} + \mu \delta_{ik} \delta_{jm} + \gamma \delta_{im} \delta_{jk}) \frac{\partial u_k}{\partial x_m}$$

$$= \alpha \delta_{ij} \frac{\partial u_k}{\partial x_k} + \mu \frac{\partial u_i}{\partial x_j} + \gamma \frac{\partial u_j}{\partial x_i}$$

As a consequence of Lemma 3.2, S is symmetric. Therefore, $\gamma = \mu$ and we obtain

$$
\begin{aligned}
\sigma_{ij} &= \alpha\delta_{ij}\frac{\partial u_k}{\partial x_k} + \mu\left(\frac{\partial u_i}{\partial x_j} + \frac{\partial u_j}{\partial x_i}\right) \\
&= \alpha\delta_{ij}e_{kk} + 2\mu e_{ij} \qquad \text{using the deformation tensor (3.15)} \\
&= \alpha\delta_{ij}\left(\sum_k e_{kk}\right) + 2\mu e_{ij} \qquad \text{without summation convention} \qquad (3.17)
\end{aligned}
$$

We calculate its trace[2]

$$
\sum_{i=1}^{3}\sigma_{ii} = \alpha\left(\sum_{i=1}^{3}\sum_{k=1}^{3}e_{kk}\right) + 2\mu\sum_{i=1}^{3}e_{ii} = 3\alpha\left(\sum_{k=1}^{3}e_{kk}\right) + 2\mu\sum_{i=1}^{3}e_{ii}
$$

$$
= \left(\sum_{k=1}^{3}e_{kk}\right)(3\alpha + 2\mu)
$$

According to (3.8), we have $\sum_i \sigma_{ii} = 0$ and therefore $\alpha = -\frac{2}{3}\mu$. After substituting into (3.17), we get

$$
\begin{aligned}
\sigma_{ij} &= 2\mu\left(e_{ij} - \frac{1}{3}\delta_{ij}\sum_k e_{kk}\right) \\
&= \mu\left(\frac{\partial u_i}{\partial x_j} + \frac{\partial u_j}{\partial x_i} - \frac{2}{3}\delta_{ij}\sum_k \frac{\partial u_k}{\partial x_k}\right) \qquad \text{by (3.15)}
\end{aligned}
$$

By substituting into (3.7), we obtain for the stress tensor

$$
T_{ij} = -p\,\delta_{ij} + \mu\left(\frac{\partial u_i}{\partial x_j} + \frac{\partial u_j}{\partial x_i} - \frac{2}{3}\delta_{ij}\sum_k \frac{\partial u_k}{\partial x_k}\right) \qquad (3.18)
$$

> The parameter μ is a temperature-dependent material constant and is referred to as *dynamic viscosity*. It can be converted into the *kinematic viscosity* ν using the relation
>
> $$
> \mu = \nu\varrho \qquad (3.19)
> $$

---

[2] The *trace* of a matrix $S = (\sigma_{ij})$ is the sum of its diagonal elements $\operatorname{tr} S = \sum_{i=1}^{3}\sigma_{ii}$.

Table 3.1 Dynamic viscosity of selected substances.

Substance	μ (mPa s)
Air (20° C)	0.018
Petroleum (20° C)	0.65
Water (5° C)	1.52
Water (20° C)	1
Motor oil (150° C)	3
Blood (37° C)	3–4 (in large vessels)
Motor oil (25° C)	100
Honey (20° C)	10,000
Bitumen (depending on type)	10^7–10^{14}

Viscosity is a measure of a fluid's resistance to flow (Table 3.1). Corresponding values can even be measured for materials such as bitumen and asphalt. However, for modeling through Navier-Stokes equations we need the aforementioned Assumption 1, which is not automatically fulfilled. For the latter, so-called *non-Newtonian fluids* such as blood, calculations based on the Navier-Stokes equations can yield inaccurate results.

3.4 The Navier-Stokes Equations

Now we can summarise the results. Let's start with the boundary conditions. Experiments with the addition of dye in flows show that the velocity near solid boundaries comes to a standstill. This justifies the boundary condition $\mathbf{u} = 0$. Chapter 5 shows that it is essential for the generation of vorticity.

Theorem 3.1 *An incompressible flow in the domain Ω satisfies the Navier-Stokes equations*

$$\frac{\partial \mathbf{u}}{\partial t} + (\mathbf{u} \cdot \nabla)\mathbf{u} + \frac{1}{\varrho}\nabla p + \frac{1}{\varrho}\mathbf{f} = \nu\Delta\mathbf{u} \qquad \text{in } \Omega \qquad (3.20)$$

$$\operatorname{div}\mathbf{u} = 0 \qquad \text{in } \Omega \qquad (3.21)$$

$$\mathbf{u} = 0 \qquad \text{on } \partial\Omega \qquad (3.22)$$

Proof Equation (3.21) results from Corollary 2.3 for incompressible flows. Considering (3.21), we can simplify (3.18) to

$$T_{ij} = -p\,\delta_{ij} + \mu\left(\frac{\partial u_i}{\partial x_j} + \frac{\partial u_j}{\partial x_i}\right)$$

Taking into account the summation convention, we evaluate its divergence

$$\frac{\partial T_{ij}}{\partial x_j} = -\frac{\partial p}{\partial x_j}\,\delta_{ij} + \mu\left(\frac{\partial^2 u_i}{\partial x_j^2} + \frac{\partial}{\partial x_i}\frac{\partial u_j}{\partial x_j}\right)$$

$$= -\frac{\partial p}{\partial x_i} + \mu\frac{\partial^2 u_i}{\partial x_j^2} \qquad \text{by (3.21)}$$

$$\text{or} \quad \operatorname{div}\mathsf{T} = -\nabla p + \varrho\nu\nabla^2\mathbf{u} \qquad \text{by (3.19)}$$

After inserting into (3.4), we get

$$\varrho\frac{D\mathbf{u}}{Dt} = \mathbf{f} - \nabla p + \varrho\nu\nabla^2\mathbf{u},$$

from which we derive (3.20) by rearranging with (2.2). $\square$

Usually, the problem is formulated in a flow domain with given initial conditions $\mathbf{u}(\mathbf{x}, 0)$ and known external force density $\mathbf{f} = \mathbf{f}(\mathbf{x}, t)$, where we are interested in finding pressure $p = p(\mathbf{x}, t)$ and velocity $\mathbf{u} = \mathbf{u}(\mathbf{x}, t)$. In mathematical research, the density ϱ is often assumed to be constant. Although the existence of solutions at certain time intervals has been demonstrated, it is unknown whether there exists a smooth solution for all times in three-dimensional space.

Wind tunnel simulations are useful because flows can exhibit similar behavior despite different extents, speeds, and particle types. We want to transform the Navier-Stokes equations into a form that makes this similarity recognisable. For simplicity, we restrict ourselves to the case of vanishing external forces $\mathbf{f} = 0$. Using the variable transformation

$$\mathbf{x}' = \frac{\mathbf{x}}{L}, \qquad \mathbf{u}' = \frac{\mathbf{u}}{U}, \qquad p' = \frac{p}{\varrho U^2}, \qquad t' = \frac{tU}{L}, \tag{3.23}$$

we denote the new differential operators by ∇' and Δ'.

Problem 3.2 Use (3.23) to transform the Navier-Stokes equations (3.20), (3.21) for $\mathbf{f} = 0$ into the following form.

$$\frac{\partial\mathbf{u}'}{\partial t'} + (\mathbf{u}' \cdot \nabla')\mathbf{u}' + \nabla'p' = \frac{\nu}{UL}\Delta'\mathbf{u}' \qquad \text{in } \Omega \tag{3.24}$$

$$\operatorname{div}\mathbf{u}' = 0 \qquad \text{in } \Omega \tag{3.25}$$

After substituting $\mathrm{Re} = UL/\nu$ and dropping the primes, we obtain the *Navier-Stokes equations in dimensionless variables.*

$$\frac{\partial \mathbf{u}}{\partial t} + (\mathbf{u} \cdot \nabla)\mathbf{u} + \nabla p = \frac{1}{\mathrm{Re}}\Delta \mathbf{u} \qquad \text{in } \Omega \tag{3.26}$$

$$\mathrm{div}\,\mathbf{u} = 0 \qquad \text{in } \Omega \tag{3.27}$$

$$\mathbf{u} = 0 \qquad \text{on } \partial\Omega \tag{3.28}$$

With a suitable choice of U and L, the flow can be characterized by a single parameter.

The constant

$$\mathrm{Re} = \frac{UL}{\nu} \tag{3.29}$$

is called the *Reynolds number.*

The choice of L and U is somewhat arbitrary and should be made with caution. Usually, L is chosen as the length of an obstacle, e.g. of an insect or aeroplane. For U, we can use the undisturbed flow velocity at a sufficient distance from the obstacle. Some examples are listed in Table 3.2. Fast flows of weakly viscous fluids are usually characterised by high Reynolds numbers.

Table 3.2 Typical values for Reynolds numbers at $\nu = 0.15\,\mathrm{cm}^2/\mathrm{s}$ for air and $\nu = 0.01\,\mathrm{cm}^2/\mathrm{s}$ for water. Data: [2], [3].

	U (cm/s)	L (cm)	$\mathrm{Re} = UL/\nu$
Swimming bacterium[a]	10^{-2}	10^{-5}	10^{-5}
Swimming sperm	10^{-2}	10^{-3}	10^{-3}
Ciliate (Ciliophora)[b]	10^{-2}	10^{-1}	10^{-1}
Minnow	5	2	10^3
Small wasp[c]	0.06	10^2	15
Locust[d]	4	400	10^4
Pigeon[e]	25	100–1000	10^5
Medium fish	50	100	$5 \cdot 10^4$
Car on motorway	300	300	10^6
Passenger aeroplane	3000	3000	10^8

[a] Lower limit of the Navier-Stokes theory due to the influence of Brownian motion

[b] Flagellum length $\approx 10^{-3}\,\mathrm{cm}$

[c] U is the speed of the wing tip during hovering

[d] For wing flow, $\mathrm{Re} \approx 2000$ would apply

[e] For wing flow, $\mathrm{Re} \approx 10^4$ would apply

A comparison of the Reynolds numbers gives an idea of the challenges other creatures face. The following example is taken from [3].

Example 3.1 To illustrate the living conditions of a ciliate, we imagine a swimmer of size $L = 2\,$m at the speed $U = 1\,$m/s in the environment of the ciliate (Re $= 10^{-1}$). The man would be in a fluid of kinematic viscosity $v = UL/\mathrm{Re} = 2 \cdot 10^5\,\mathrm{cm}^2/\mathrm{s}$. This value corresponds to a dynamic viscosity of hot tar or bitumen from Table 3.1, as can be checked after conversion using (3.19) and the tar density $\varrho = 1.175\,\mathrm{g/cm}^3$. The comparison illustrates the role of frictional forces in the locomotion of single-celled organisms.

The specific flow characteristics of their environment have forced the various organisms to develop an adapted body structure and suitable methods of locomotion.

At the same time, the calculation of the flow can be simplified if we keep only essential components in the equations and neglect others. In modeling for creatures, a distinction is made between the cases Re $\ll 1$ (microorganisms) and Re $\gg 1$ (fish, birds and insects).

In the context of flow around obstacles, we mention some common terms.

The *inflow* refers to the flow or fluid as it moves towards the obstacle. It is often modeled by assuming a constant velocity, the so-called *inflow velocity* **U**.

The *freestream velocity* $\mathbf{u}_\infty$ refers to the constant velocity of fluid when it flows far from solid boundaries where it could be deflected, slowed down, accelerated, or compressed. In many situations, a flow returns to its inflow velocity at a sufficient distance after passing an obstacle.

The *wake* is the flow behind the obstacle. It is mainly determined by the shape of the obstacle.

A flow exerts a force on an obstacle, which can be decomposed as follows. The component that acts
- in the opposite direction to the inflow is referred to as *drag*.
- in the direction perpendicular to the inflow is called *lift*.

3.5   Laminar and Turbulent Flows

Towards the end of the nineteenth century, Reynolds conducted experiments with flows in pipes, injecting dye through an opening. At lower flow velocities, the dye was transported in a thread. As the velocity increased, the thread began to fray and form spots, with the dye quickly filling the entire pipe. Such behaviour is called the transition from a laminar to a turbulent flow (Fig. 3.2).

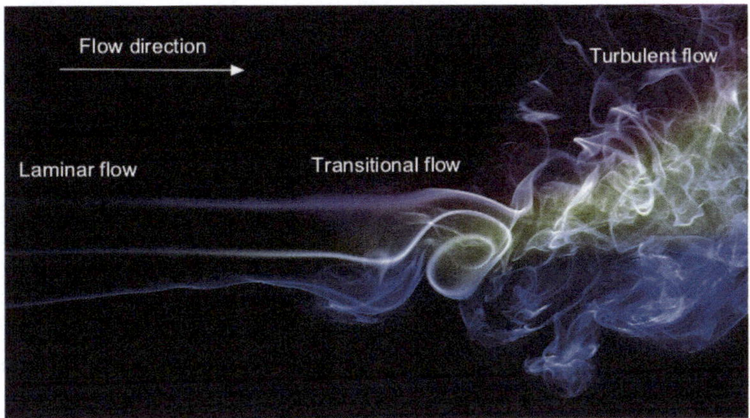

Fig. 3.2 In laminar flow, the fluid flows in layers that do not mix. Here, steady flows are possible. In contrast, a turbulent flow is characterised by vorticity at various scales and mixing. The transition from laminar to turbulent flow occurs over a certain distance, which depends particularly on the local pressure gradient. Author: airguy1988, CC BY-ND (Attribution NoDerivatives) license. https://creativecommons.org/licenses/by-nd/4.0/.

Turbulent flows require friction and exhibit stronger shear stresses than laminar ones. They form a multitude of eddies and wirls that interact and—in contrast to the laminar case—cause random fluctuations of the particles. Thus, turbulent flows show a significantly different behaviour than most cases studied in this book. Turbulence is always unsteady (Sect. A.5) and difficult to calculate due to the structure of its vorticity. They influence the boundary behaviour of many flows and are indispensable for explaining aerodynamic lift. Transitions between laminar and turbulent flows are possible in both directions.

In the study of turbulence, it turns out that the velocity is not the sole characteristic for its occurrence. Although turbulent flows can also be observed at low velocities, a high Reynolds number $\mathrm{Re} = UL/\nu$ serves as a clue in many practical cases. Thus, higher velocities U as well as smaller length scales L and viscosities ν contribute to the formation of turbulence.

Owing to the size and speed of cars, aeroplanes and rockets, turbulence effects play an important role in engineering. With corresponding energy intake in the boundary layer, fuel costs increase. Therefore, a reduction in turbulence is often desirable.

For certain processes, turbulence is indispensable, for example for generating a starting vortex on the wing (Sect. 5.6). In the context with the boundary layer thickness, it becomes understandable that turbulent boundary layers can adhere more firmly to surfaces than laminar ones (see Sects. 5.4, 5.5 and [6]). This reduces their turbulent area in the wake, leading to a reduction in energy losses. These considerations must be included in the energy balance at high speeds.

3.6   The Inviscid-Flow Approximation

For $\nu \Delta \mathbf{u} = 0$, the system (3.20), (3.21), (3.22) simplifies to the *Euler equations*

$$\frac{\partial \mathbf{u}}{\partial t} + (\mathbf{u} \cdot \nabla)\mathbf{u} + \frac{1}{\varrho}\nabla p + \frac{1}{\varrho}\mathbf{f} = 0 \quad \text{in } \Omega \tag{3.30}$$

$$\operatorname{div} \mathbf{u} = 0 \quad \text{in } \Omega \tag{3.31}$$

$$\mathbf{u} \cdot \mathbf{n} = 0 \quad \text{on } \partial\Omega \tag{3.32}$$

This approximation is also possible for $\nu \neq 0$, see Exercise 4.5. Equation (3.30) is satisfied for an ideal fluid (Sect. 3.2), whose stress tensor only contains the pressure component. By (3.31), the fluid is incompressible.

Condition (3.32) means that the fluid cannot penetrate the fixed boundary. However, sliding along the boundary is possible, i.e. the velocity may have a tangential component at the boundary. Maintaining the stronger requirement (3.22) would be too restrictive in the inviscid model, as a number of important use cases would only have the trivial solution $\mathbf{u} = 0$.

At certain boundary points, the velocity can disappear permanently even in the inviscid model (see Fig. 3.3).

A point $\mathbf{x}$ with $\mathbf{u}(\mathbf{x}, t) = 0$ is called a *stagnation point*. Thus, the boundary consists of streamlines, which are separated from each other by stagnation points.

If friction is neglected, the performance of work on the flowing particle causes a change in mechanical energy according to the law of conservation of energy.

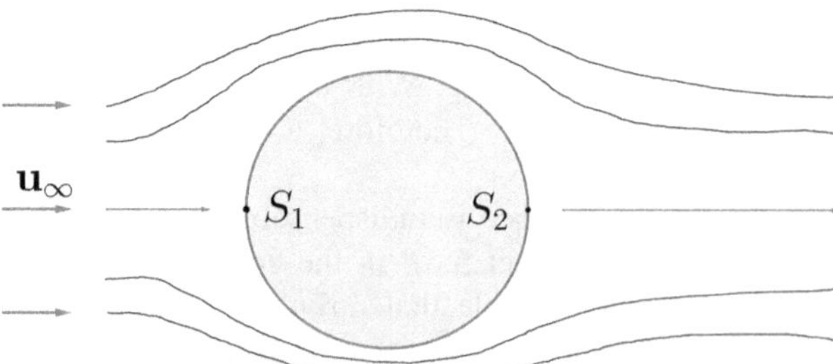

Fig. 3.3 For uniform flow around a circular obstacle, stagnation points are at S_1 and S_2. The upper and lower arcs each form a streamline.

The following statement was discovered in the eighteenth century by the brothers Daniel and Johann Bernoulli.

Theorem 3.2 (Bernoulli) *In the region Ω we consider an ideal fluid with constant density in a stationary flow, where the external force is conservative, i.e. it has a potential Φ. Then it holds*

$$\frac{p}{\varrho} + \Phi + \frac{1}{2}\|\mathbf{u}\|^2 = \text{const} \qquad \text{on each streamline} \tag{3.33}$$

Under the additional assumption of a vortex-free flow,

$$\frac{p}{\varrho} + \Phi + \frac{1}{2}\|\mathbf{u}\|^2 = \text{const} \qquad \text{in } \Omega \tag{3.34}$$

Proof

Ad. (3.33):   Since ϱ is constant and $\mathbf{f}$ is conservative, we find a potential Φ with $\frac{1}{\varrho}\mathbf{f} = \nabla\Phi$. From (3.30), it follows that

$$(\mathbf{u} \cdot \nabla)\mathbf{u} + \nabla\left(\frac{p}{\varrho}\right) + \nabla\Phi = 0 \tag{3.35}$$

Furthermore, we obtain from (1.17) for $\mathbf{a} = \mathbf{b} = \mathbf{u}$

$$\nabla\left(\|\mathbf{u}\|^2\right) = 2(\mathbf{u} \cdot \nabla)\,\mathbf{u} + 2\mathbf{u} \times \text{rot}\,\mathbf{u}$$

$$(\mathbf{u} \cdot \nabla)\,\mathbf{u} = \frac{1}{2}\nabla\left(\|\mathbf{u}\|^2\right) - \mathbf{u} \times \text{rot}\,\mathbf{u} \tag{3.36}$$

and after inserting into (3.35)

$$\nabla\left(\frac{p}{\varrho} + \Phi + \frac{1}{2}(\|\mathbf{u}\|^2)\right) = \mathbf{u} \times \text{rot}\,\mathbf{u} \tag{3.37}$$

If we form the scalar product with $\mathbf{u}$ on both sides and consider that $\mathbf{u} \times \text{rot}\,\mathbf{u}$ is perpendicular to $\mathbf{u}$, it follows that

$$\mathbf{u} \cdot \nabla\left(\frac{p}{\varrho} + \Phi + \frac{1}{2}(\|\mathbf{u}\|^2)\right) = 0$$

Thus, the directional derivative of

$$H := \frac{p}{\varrho} + \Phi + \frac{1}{2}(\|\mathbf{u}\|^2)$$

along $\mathbf{u}$ disappears. Consequently, H remains constant on each integral curve of $\mathbf{u}$, i.e. on each streamline.

Ad. (3.34):   We look again at (3.37), where the right-hand side vanishes for a vortex-free field in the entire flow region Ω. Thus, H remains constant in Ω. $\square$

Example 3.2 In medicine, *aortic valve stenosis* refers to the narrowing of the aortic valve, which can occur in the left or right ventricle (see Figs. 3.4 and 3.5). Stenoses increase the load on the heart muscle, which can lead to heart failure.

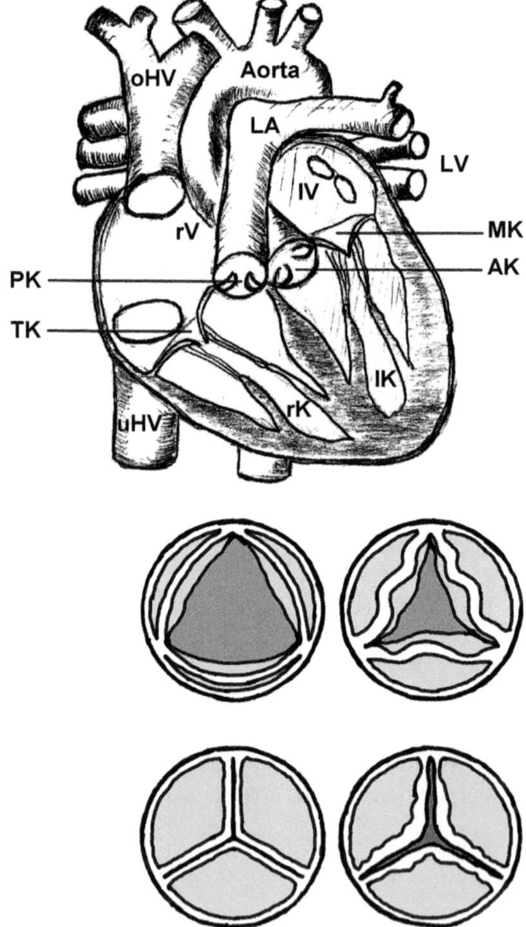

Fig. 3.4 Diagram of the heart: *oHV* upper vena cava, *uHV* lower vena cava, *LA* pulmonary artery, *LV* pulmonary vein, *rA* right atrium, *lA* left atrium, *rV* right ventricle, *lV* left ventricle, *PK* pulmonary valve, *TK* tricuspid valve, *MK* mitral valve, *AK* aortic valve.

Fig. 3.5 Normal aortic valve (open at the top left, closed at the bottom left) and aortic stenosis (open at the top right, closed at the bottom right).

Table 3.3 Pressure differences at the aortic valve between the aorta and the left ventricle. Data: [4].

Severity	Average pressure difference Δp (mm Hg)
Normal	<5
Mild stenosis	5–25
Moderate stenosis	25–50
Severe stenosis	>50

The pressure difference can be measured by inserting a pressure catheter. As a non-invasive alternative, the maximum velocity of erythrocytes (blood flow velocity) is measured using echocardiographic methods. This allows the pressure difference to be calculated.

We denote the values of pressure, velocity and potential energy before the narrowing by p_1, $\mathbf{u}_1$, Φ_1 and after the narrowing by p_2, $\mathbf{u}_2$, Φ_2. From Bernoulli's theorem we get

$$p_1 + \varrho\Phi_1 + \frac{\varrho}{2}\|\mathbf{u}_1\|^2 = p_2 + \varrho\Phi_2 + \frac{\varrho}{2}\|\mathbf{u}_2\|^2$$

In the considered section, the potential energy remains almost constant, i.e. $\Phi_1 = \Phi_2$. Using $\Delta p = p_1 - p_2$ and the approximate value $\varrho \approx 8$ for blood, we find that

$$\Delta p = 4(u_2^2 - u_1^2)$$

In most clinical cases, the velocity before the narrowing $\mathbf{u}_1 < 1$ can be neglected compared to $\mathbf{u}_2$. Then, the pressure difference can be calculated with

$$\Delta p = 4u_2^2 \tag{3.38}$$

For example, if we measure the flow velocity $u_2 = 4\,\text{m/s}$ at the aortic valve of the left ventricle, we obtain the pressure difference $\Delta p = 64\,\text{mm Hg}$, indicating severe stenosis (Table 3.3). It should be noted that the method based on the Doppler effect records the highest speed and thus overestimates pressure differences.

Example 3.3 For centuries, an energy-saving cooling system known as *Badgir* [3] has been used in the Middle East. It allows air circulation from the open tower top to an underground channel through a massive tower that extends down to the basement (see Fig. 3.6).

We denote the wind speed and pressure at the tower top by $\mathbf{u}(\mathbf{x}_t)$, $p(\mathbf{x}_t)$ and at the entrance shaft to the channel by $\mathbf{u}(\mathbf{x}_e)$, $p(\mathbf{x}_e)$.

---

[3] Persian: wind catcher.

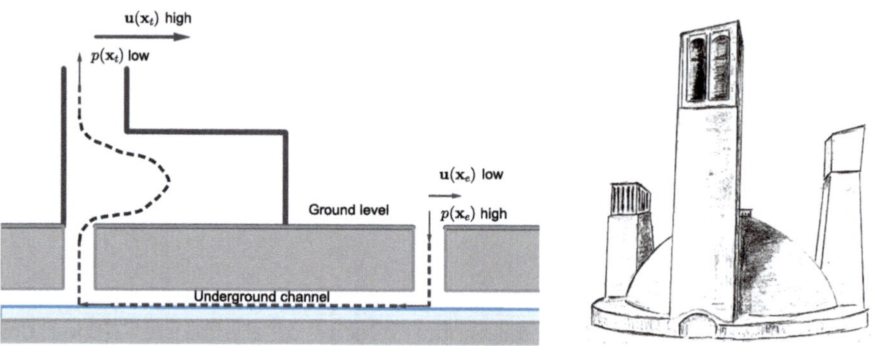

Fig. 3.6 (left) Working scheme of a Badgir. (right) Badgir as a water reservoir in Iran.

If we neglect the potential energy Φ of the air particles, Bernoulli's theorem gives

$$\frac{p(\mathbf{x}_t)}{\varrho} + \frac{1}{2}\|\mathbf{u}(\mathbf{x}_t)\|^2 = \frac{p(\mathbf{x}_e)}{\varrho} + \frac{1}{2}\|\mathbf{u}(\mathbf{x}_e)\|^2$$

At height, the wind speed is greater than near the ground. Consequently, the pressure at height is lower than near the ground and the air can escape more easily from the top of the tower than from the entrance shaft. This creates air circulation from the entrance shaft through the channel into the building, which is represented by a dashed line in Fig. 3.6. As the air in the underground channel sweeps over the water, it cools down and absorbs water vapour, which is subsequently absorbed by the clay walls in the building. The subsequent evaporation removes heat from the room air, providing additional cooling. At outside temperatures of up to 48°C, a temperature reduction of up to 20°C can be achieved in the building.

Finally, we examine an important special case of inviscid flows.

Lemma 3.4 *A vortex-free, incompressible flow* $\mathbf{u}$ *satisfies the Euler's equation of motion (3.30).*

Proof From (1.14), it follows that

$$\Delta\mathbf{u} = \nabla \underbrace{(\nabla \cdot \mathbf{u})}_{=0} - \nabla \times \underbrace{(\nabla \times \mathbf{u})}_{=0} = 0$$

Therefore, the term $\nu\Delta\mathbf{u}$ in the Navier-Stokes equation (3.20) vanishes and it reduces to (3.30). $\qquad\square$

Thus, vortex-free, incompressible flows can be considered as inviscid. In the following chapters, we use the Navier-Stokes equations (3.20), (3.21), (3.22) or their inviscid-flow simplification (3.30), (3.31), (3.32) to solve various problems. Since the gravity of liquid particles plays a role, we usually assume $\mathbf{f} \neq 0$ in hydrodynamic problems. On the other hand, the weight of gas particles can often be neglected, which justifies the assumption of the external force $\mathbf{f} = 0$ in aerodynamics.

3.7 Propagation of Vortex Fields

So far, we have focused on vortex-free flows. However, as we will learn in Chap. 5, vorticity is indispensable for aerodynamic lift. They are essentially created directly at the wings and are carried along with the flow. While the generation of vorticity at increasing distance from the wing does not play a role for the lift, their downstream transport is important.

Since friction can be neglected outside a thin layer around the wing, we model the more distant airspace using the Euler equations (3.30), (3.31), (3.32) in the form

$$\frac{\partial \mathbf{u}}{\partial t} + (\mathbf{u} \cdot \nabla)\mathbf{u} + \frac{1}{\varrho}\nabla p = 0 \tag{3.39}$$

$$\operatorname{div} \mathbf{u} = 0 \tag{3.40}$$

Lemma 3.5 *Consider an aerodynamic flow with constant density ϱ determined by (3.39) and (3.40). Then, the vortex transport is described by*

$$\frac{D\omega}{Dt} = (\boldsymbol{\omega} \cdot \nabla)\mathbf{u} \tag{3.41}$$

For two-dimensional flows, it holds

$$\frac{D\omega}{Dt} = 0 \tag{3.42}$$

Proof From (1.17) we obtain for $\mathbf{a} = \mathbf{b} = \mathbf{u}$ after rearrangement

$$(\mathbf{u} \cdot \nabla)\,\mathbf{u} = \frac{1}{2}\nabla\,(\mathbf{u} \cdot \mathbf{u}) - \mathbf{u} \times \operatorname{rot}\mathbf{u} \tag{3.43}$$

If we apply the curl operator to (3.39), we get

$$\frac{\partial \omega}{\partial t} + \operatorname{rot}(\mathbf{u} \cdot \nabla)\mathbf{u} = 0 \qquad \text{by (1.19)}$$

$$\frac{\partial \omega}{\partial t} + \operatorname{rot}\left[\frac{1}{2}\nabla\,(\mathbf{u} \cdot \mathbf{u}) - \mathbf{u} \times \operatorname{rot}\mathbf{u}\right] = 0 \qquad \text{by (3.43)}$$

$$\frac{\partial \omega}{\partial t} - \operatorname{rot}\left[\mathbf{u} \times \operatorname{rot}\mathbf{u}\right] = 0 \qquad \text{by (1.19)} \tag{3.44}$$

Using (1.16), we find that

$$\operatorname{rot}(\mathbf{u} \times \operatorname{rot}\mathbf{u}) = (\operatorname{rot}\mathbf{u} \cdot \nabla)\mathbf{u} - (\mathbf{u} \cdot \nabla)\operatorname{rot}\mathbf{u} + \mathbf{u}\,(\operatorname{div}\operatorname{rot}\mathbf{u}) - \operatorname{rot}\mathbf{u}\,\operatorname{div}\mathbf{u}$$

$$= (\boldsymbol{\omega} \cdot \nabla)\mathbf{u} - (\mathbf{u} \cdot \nabla)\boldsymbol{\omega} \qquad \text{by (1.18), (3.40)}$$

By substitution into (3.44) we obtain

$$\frac{\partial\boldsymbol{\omega}}{\partial t} - (\boldsymbol{\omega} \cdot \nabla)\mathbf{u} + (\mathbf{u} \cdot \nabla)\boldsymbol{\omega} = 0$$

$$\frac{D\boldsymbol{\omega}}{Dt} - (\boldsymbol{\omega} \cdot \nabla)\mathbf{u} = 0$$

According to (1.20), the vorticity in the two-dimensional case is perpendicular to the plane of the velocity field.

$$\mathbf{u} = \begin{pmatrix} u(x,y) \\ v(x,y) \\ 0 \end{pmatrix}, \qquad \boldsymbol{\omega} = \begin{pmatrix} 0 \\ 0 \\ \omega_3 \end{pmatrix}$$

Using (1.12), we obtain the directional derivative

$$(\boldsymbol{\omega} \cdot \nabla)\mathbf{u} = \begin{pmatrix} \omega_3\,\partial_z u \\ \omega_3\,\partial_z v \\ \omega_3\,\partial_z 0 \end{pmatrix} = \begin{pmatrix} 0 \\ 0 \\ 0 \end{pmatrix}$$

and thus (3.42). $\qquad\qquad\square$

In the two-dimensional model, vorticity corresponds to a scalar quantity. According to (3.42), it does not change along a trajectory and remains constant for each particle. (Remind that a particle consists of a group of neighbouring molecules.)

In the inviscid model, vorticity is transported by the flow, with the Jacobian matrix (2.12) causing its distortion (see Fig. 3.7). To illustrate this, we consider a fluid particle that starts at time zero at position $\mathbf{a}$ and reaches $\mathbf{x} = \mathbf{x}(\mathbf{a}, t)$ at time t.

Lemma 3.6 *Let $M_t : \mathbf{a} \mapsto \mathbf{x}(\mathbf{a}, t)$ be the Lagrange mapping and DM_t its Jacobian matrix. Then the vorticity is given by*

$$\boldsymbol{\omega}(\mathbf{x}(\mathbf{a}, t), t) = DM_t\,\boldsymbol{\omega}(\mathbf{a}, 0) \tag{3.45}$$

Proof By

$$\hat{\boldsymbol{\omega}}(\mathbf{a}, t) := \boldsymbol{\omega}(\mathbf{x}(\mathbf{a}, t), t)$$

we describe the vortex field depending on $\mathbf{a}$ and t. Along a fixed trajectory, the vorticity is determined solely by the time t. Therefore, the material derivative D/Dt along the flow simplifies to the time derivative, i.e. we get

$$\frac{\partial}{\partial t}\hat{\omega}(\mathbf{a}, t) = \frac{D}{Dt}\omega(\mathbf{x}(\mathbf{a}, t), t)$$

and by (3.41)

$$\frac{\partial\hat{\omega}}{\partial t} = (\hat{\omega} \cdot \nabla)\mathbf{u}$$

We define the vector field $\boldsymbol{\xi} = \boldsymbol{\xi}(\mathbf{a}, t)$ by

$$\boldsymbol{\xi}(\mathbf{a}, t) = DM_t\,\omega(\mathbf{a}, 0) \tag{3.46}$$

Using (2.12) and the notation

$$DM_t = \begin{pmatrix} \frac{\partial x_1}{\partial a_1} & \frac{\partial x_1}{\partial a_2} & \frac{\partial x_1}{\partial a_3} \\ \frac{\partial x_2}{\partial a_1} & \frac{\partial x_2}{\partial a_2} & \frac{\partial x_2}{\partial a_3} \\ \frac{\partial x_3}{\partial a_1} & \frac{\partial x_3}{\partial a_2} & \frac{\partial x_3}{\partial a_3} \end{pmatrix} = \frac{\partial \mathbf{x}}{\partial \mathbf{a}},$$

we calculate

$$\frac{\partial DM_t}{\partial t} = \frac{\partial}{\partial t}\frac{\partial \mathbf{x}}{\partial \mathbf{a}} = \frac{\partial}{\partial \mathbf{a}}\frac{\partial \mathbf{x}}{\partial t} = \frac{\partial \mathbf{u}}{\partial \mathbf{a}} = \frac{\partial \mathbf{u}}{\partial \mathbf{x}}\frac{\partial \mathbf{x}}{\partial \mathbf{a}},$$

applying the chain rule in the last step. Then,

$$\frac{\partial \boldsymbol{\xi}}{\partial t} = \frac{\partial DM_t}{\partial t}\omega(\mathbf{a}, 0) = \frac{\partial \mathbf{u}}{\partial \mathbf{x}}\frac{\partial \mathbf{x}}{\partial \mathbf{a}}\omega(\mathbf{a}, 0) = \frac{\partial \mathbf{u}}{\partial \mathbf{x}}\boldsymbol{\xi} \qquad \text{by (3.46)}$$

$$= (\boldsymbol{\xi} \cdot \nabla)\mathbf{u}$$

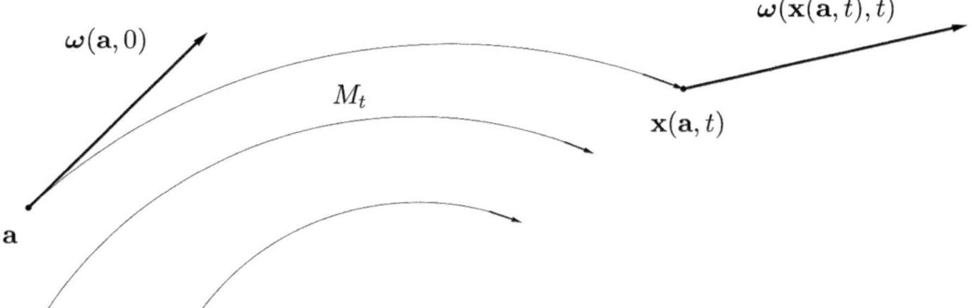

Fig. 3.7 The Lagrange mapping M_t transports the vortex field.

We have thus shown that the vector fields $\hat{\omega}(\mathbf{a}, t)$ and $\boldsymbol{\xi}(\mathbf{a}, t)$ satisfy the same differential equation. Since $DM_0 = \mathrm{Id}$, they also agree at the start time $t = 0$ and are thus identical. $\qquad\square$

When friction is taken into account, diffusion effects come into play.

Problem 3.3 Prove for the diffusion term of the vorticity that

$$\mathrm{rot}\,\Delta\mathbf{u} = \Delta\omega \tag{3.47}$$

In analogy to Lemma 3.5, we can derive the *advection-diffusion equation of vorticity* from the Navier-Stokes equations with constant density.

$$\frac{D\omega}{Dt} = (\boldsymbol{\omega} \cdot \nabla)\mathbf{u} + \nu\Delta\omega \tag{3.48}$$

It shows similarities to the heat conduction equation $\frac{\partial T}{\partial t} = k\,\Delta T$. ($k$ denotes the thermal conductivity.)

References and Further Reading

1. Aris R.: *Vectors, Tensors and the Basic Equations of Fluid Mechanics.* Prentice-Hall, New Jersey (1962)
2. Childress S.: *An Introduction to Theoretical Fluid Mechanics.* Courant Lecture Notes in Mathematics Vol. 19, AMS, Providence RI (2009)
3. Childress S.: *Mechanics of swimming and flying.* Cambridge University Press, Cambridge (1981)
4. Gross K., Bejot T., Shernan S.: *eecho.* https://www.e-echocardiography.com/normal-values Accessed 3rd Sept 2024
5. Hodge P.G.: *On Isotropic Cartesian Tensors.* The American Mathematical Monthly (1961), 68(8), 793-795.
6. Platzer M.F., Jones K.D.: *Concepts of aero- and hydromechanics.* WIT, Flow Phenomena in Nature, Vol. 1 (2006), pp. 5-19

Aerodynamic Lift 4

The foundations of fluid mechanics were laid in the eighteenth century, mainly by Euler, d'Alembert and the Bernoulli brothers. Their completion to the Navier-Stokes equations took place until the first half of the nineteenth century, with Navier, Poisson and Stokes making corresponding contributions. But it was not until the beginning of the twentieth century that Kutta, Joukowski and Prandtl succeeded in understanding aerodynamic lift. One reason for the delay was the difficulty in finding useful simplifications of the Navier-Stokes equations. At first glance, a variety of approaches appeared, but after careful examination, most of them lost their practical value. The flow around a thin plate with a rounded nose can be examined with approaches whose solutions correspond to the Figs. 4.1, 4.2, 4.3, and 4.4 (see [1]). All describe real situations, but only in Fig. 4.3 is stable flight possible. [1] Recognising the suitable model required a sharp observation of the flow around the wing.

The diversity of the flows Figs. 4.1, 4.2, 4.3, and 4.4 is largely determined by the behaviour within a narrow layer around the plate, which is called the boundary layer. Aerodynamic lift is complicated to explain because the crucial processes of friction and vortex formation take place in this layer, the formation and stability of which depend on the wing geometry and inflow. To uncover these relations, a series of step-by-step simplifications of the Navier-Stokes equations is carried out, which will be retraced in the following chapters.

Theoretical problems usually come into focus when they become practically implementable. Lilienthal's test flight with a hang glider in 1894 was preceded by extensive observations and wind tunnel measurements, which led to the recognition of the specific profile curvature as the prerequisite for lift (see Figs. 4.5 and 4.6).

---

[1] Figure 4.1 illustrates the situation immediately before takeoff and is explained in Sect. 4.6. Both situations in Figs. 4.2 and 4.4 describe a loss of flight stability and are therefore also of interest, see Sect. 5.4 and [2].

© The Author(s), under exclusive license to Springer-Verlag GmbH, DE, part of Springer Nature 2026
R. Spielmann, *Theoretical Fluid Mechanics*,
https://doi.org/10.1007/978-3-662-72240-4_4

Fig. 4.1 Flow around a plate with stagnation points near the corners. Formation of a torque, but neither lift nor drag.

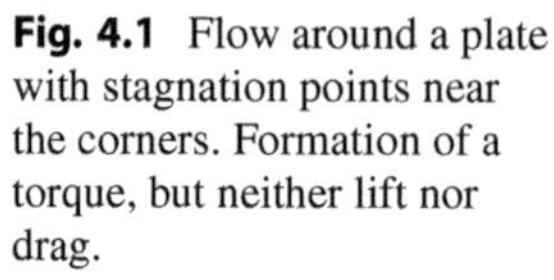

Fig. 4.2 Flow separation near the corners. Low lift, high drag (Helmholtz-Kirchhoff flow).

Fig. 4.3 Stagnation point at the nose, flow separation at the tail. High lift, no drag (Kutta-Joukowski flow).

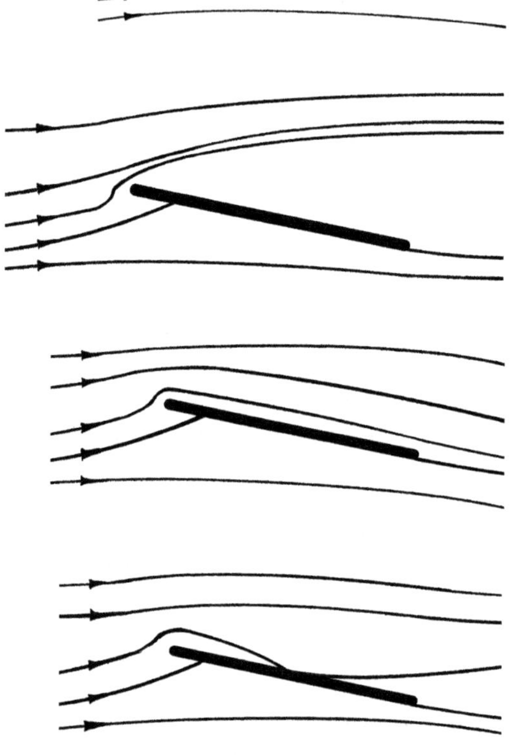

Fig. 4.4 Flow separation on the upper side with formation of a "dead water" area. Reduced lift, low drag.

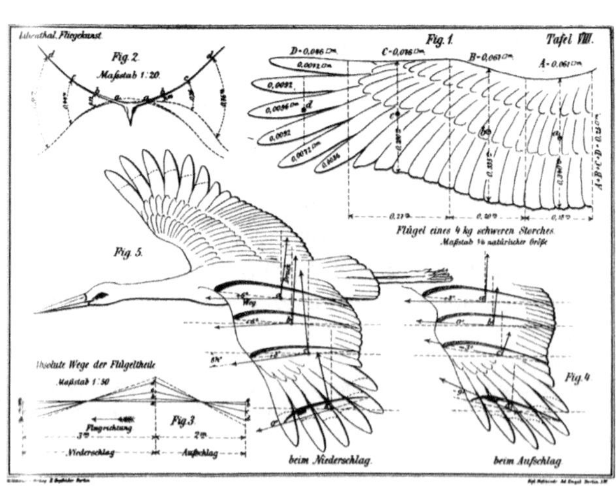

Fig. 4.5 Measurement of stork wings from Lilienthal: *Bird flight as the basis of aviation*, 1889.

Fig. 4.6 Lilienthal's biplane in October 1895. One year later, the aviation pioneer died in an accident.

Fig. 4.7 In 1910, the Wright brothers developed the Model B. It was the first aircraft to be used for military purposes in the United States.

Joukowski acquired one of the gliding apparatuses and conducted his own experiments in a wind tunnel. The practical breakthrough in aircraft construction occurred parallel to the theoretical one. In 1903, the Wright brothers succeeded in the first motor flight (see Fig. 4.7). Formulas for the lift force were discovered by Kutta in 1902 and independently by Joukowski four years later.

4.1 Streamlines and Pressure Gradient

We examine a stationary flow in the inviscid-flow approximation (3.30), (3.31), (3.32). For the Kutta-Joukowski flow (Fig. 4.3), the direction of the lift can be derived from the curvature of the streamlines (Fig. 4.8). The idea is based on the use of a travelling coordinate system, the Frenet trihedron along a streamline. The underlying theory is compiled in Sect. A.3.

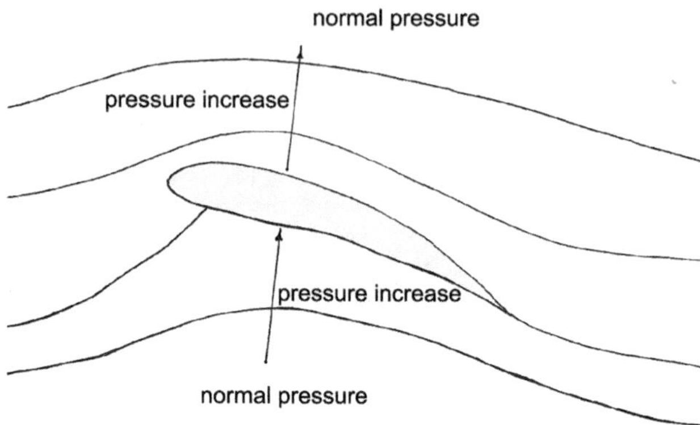

Fig. 4.8 Cross-section of a stationary flow around a wing. The pressure increases in the upward direction. Owing to the small difference in altitude, the atmospheric pressure above and below the wing is the same (normal pressure).

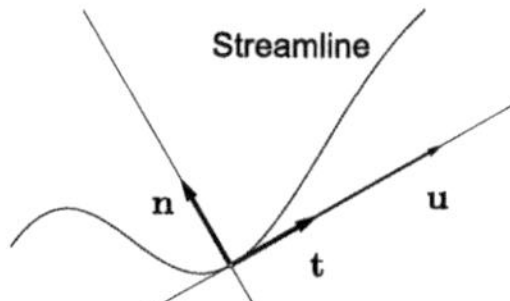

Fig. 4.9 Streamline with tangent and normal vector. The binormal vector is perpendicular to the drawing plane and points towards the viewer.

The Frenet formulas establish a relation between curvature and the change of tangent vectors, i.e., acceleration. Using the Euler equation (3.30), we are looking for a relation between streamline curvature and pressure change, which explains the lift.

First, we parametrise the section of a streamline by its arc length s. The position of the particle on it is described by $s = s(t)$. For its velocity and acceleration, we have

$$\mathbf{u} = \mathbf{u}(s(t), t), \qquad \mathbf{a} = \frac{\mathrm{d}}{\mathrm{d}t}\mathbf{u}(s(t), t) = \frac{\partial \mathbf{u}}{\partial s}\frac{\mathrm{d}s}{\mathrm{d}t} + \frac{\partial \mathbf{u}}{\partial t} \tag{4.1}$$

Now let $\{\mathbf{t}, \mathbf{n}, \mathbf{b}\}$ be an orthonormal basis at the current location of the particle, consisting of tangent, normal and binormal vector to the streamline (Fig. 4.9).

The velocity, whose magnitude we denote by $u = \|\mathbf{u}\|$, lies in the direction of the tangent vector, i.e.

$$\mathbf{u} = u\,\mathbf{t} = \frac{\mathrm{d}s}{\mathrm{d}t}\,\mathbf{t}$$

By substitution into (4.1), we get

$$\mathbf{a} = \frac{\partial(u\mathbf{t})}{\partial s}\,u + \frac{\partial \mathbf{u}}{\partial t} = (\frac{\partial u}{\partial s}\,\mathbf{t} + u\,\frac{\partial \mathbf{t}}{\partial s})\,u + \frac{\partial \mathbf{u}}{\partial t} = u\,\frac{\partial u}{\partial s}\,\mathbf{t} + u^2\,\frac{\partial \mathbf{t}}{\partial s} + \frac{\partial \mathbf{u}}{\partial t}$$

and further, taking into account the product rule $\frac{\partial(u^2)}{\partial s} = 2u\,\frac{\partial u}{\partial s}$

$$\mathbf{a} = \frac{1}{2}\,\frac{\partial(\|\mathbf{u}\|^2)}{\partial s}\,\mathbf{t} + \|\mathbf{u}\|^2\,\frac{\partial \mathbf{t}}{\partial s} + \frac{\partial \mathbf{u}}{\partial t}$$

Using the Frenet formula $\frac{\partial \mathbf{t}}{\partial s} = k(s)\,\mathbf{n}$ with streamline curvature $k(s)$, we obtain the acceleration

$$\mathbf{a} = \frac{1}{2}\,\frac{\partial(\|\mathbf{u}\|^2)}{\partial s}\,\mathbf{t} + \|\mathbf{u}\|^2\,k(s)\,\mathbf{n} + \frac{\partial \mathbf{u}}{\partial t}$$

On the other hand, $\mathbf{a}$ can be represented by the material derivative (2.4).

$$\mathbf{a} = \frac{\partial \mathbf{u}}{\partial t} + (\mathbf{u} \cdot \nabla)\mathbf{u}$$

Equating both expressions yields

$$(\mathbf{u} \cdot \nabla)\mathbf{u} = \frac{1}{2}\,\frac{\partial(\|\mathbf{u}\|^2)}{\partial s}\,\mathbf{t} + \|\mathbf{u}\|^2\,k(s)\,\mathbf{n} \tag{4.2}$$

In our orthonormal basis, the pressure gradient takes the form

$$\nabla p = \frac{\partial p}{\partial s}\,\mathbf{t} + \frac{\partial p}{\partial n}\,\mathbf{n} + \frac{\partial p}{\partial b}\,\mathbf{b}, \tag{4.3}$$

where $\frac{\partial p}{\partial s}, \frac{\partial p}{\partial n}, \frac{\partial p}{\partial b}$ denote its directional derivatives along $\mathbf{t}, \mathbf{n}, \mathbf{b}$.

If we substitute the relations (4.2), (4.3) into the Euler equation (3.30), consider the neglect of gravity $\mathbf{f} = 0$ for gas particles and limit ourselves to the stationary case $\frac{\partial \mathbf{u}}{\partial t} = 0$, we get

$$\frac{1}{2}\,\frac{\partial(\|\mathbf{u}\|^2)}{\partial s}\,\mathbf{t} + \|\mathbf{u}\|^2\,k(s)\,\mathbf{n} + \frac{1}{\varrho}\left(\frac{\partial p}{\partial s}\,\mathbf{t} + \frac{\partial p}{\partial n}\,\mathbf{n} + \frac{\partial p}{\partial b}\,\mathbf{b}\right) = 0$$

$$\left(\frac{1}{2}\,\frac{\partial(\|\mathbf{u}\|^2)}{\partial s} + \frac{1}{\varrho}\,\frac{\partial p}{\partial s}\right)\mathbf{t} + \left(\|\mathbf{u}\|^2\,k(s) + \frac{1}{\varrho}\,\frac{\partial p}{\partial n}\right)\mathbf{n} + \frac{1}{\varrho}\,\frac{\partial p}{\partial b}\,\mathbf{b} = 0 \tag{4.4}$$

Since the vectors **t**, **n**, **b** are linearly independent, their coefficients must vanish. For the component along **b**, we get

$$\frac{\partial p}{\partial b} = 0 \tag{4.5}$$

The pressure remains constant along the binormal, i.e. in the direction that would point perpendicularly out of the image towards us.
For the component of **t**, we find that

$$\frac{1}{2}\frac{\partial(\|\mathbf{u}\|^2)}{\partial s} + \frac{1}{\varrho}\frac{\partial p}{\partial s} = 0, \tag{4.6}$$

which again yields Bernoulli's theorem (Theorem 3.2). Finally, we evaluate (4.4) for the component of **n**.

Theorem 4.1 (Streamline Curvature Theorem) *For a stationary, inviscid flow it holds*

$$\|\mathbf{u}\|^2 k(s) + \frac{1}{\varrho}\frac{\partial p}{\partial n} = 0 \tag{4.7}$$

The equation allows for a surprising interpretation. As is known, the curvature $k(s) > 0$ and the vector **n** is directed towards the center K of the osculating circle. Then $\frac{\partial p}{\partial n} = -\varrho\|\mathbf{u}\|^2 k(s) < 0$, i.e. *the pressure decreases towards the center of curvature* (Fig. 4.10).

In the stable flight phase after takeoff and before landing, stationary flight is a good approximation for aircraft. The same applies to gliding flight in birds, where the wings are hardly moved. In these cases, the streamline curvature theorem allows the visualization of lift through the characteristic upwardly curved streamline (Fig. 4.8).

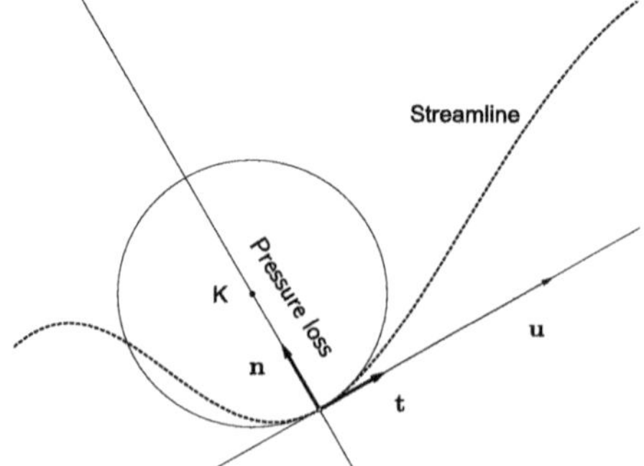

Fig. 4.10 The pressure decreases in the direction of the centre of curvature of the streamline.

The centres of curvature of the streamlines lie beneath the wing. Directly on the wing, there must be a lower pressure (negative pressure) which grows upwards and thus approaches atmospheric pressure. Under the wing, there is a higher pressure, which decreases in the downward direction and approaches atmospheric pressure. The pressure difference between both sides gives the wing lift.

Furthermore, the theorem explains some well-known natural phenomena in water and air currents. Spiral streamlines also form in *whirlpools* that occur in rivers and narrow straits due to uneven ground. In the direction of the centre of curvature, the pressure decreases, creating a suction effect.

Tornados are atmospheric air vortices with an approximately vertical axis of rotation. Usually, a tornado only lasts for minutes and rarely longer than an hour. It can develop a negative pressure of up to 100 hPa, which corresponds to about 10% of the average air pressure at sea level (= 1.013,25 hPa).

Problem 4.1 Why do umbrellas fold back and forth in rapid succession in stormy weather?

The following relation between pressure gradient and particle acceleration can also be observed in situations with friction.

Problem 4.2 Prove for a stationary, inviscid flow that particles along a streamline

(a)   are accelerated if the pressure decreases in the direction of flow,
(b)   are decelerated if the pressure increases in the direction of flow.

4.2   Potential Flows

In the previous chapter, we recognised lift using the Euler equations in streamline form. Now it is to be calculated. Since the friction term significantly complicates the Navier-Stokes equations, a solution to the lift problem using the Euler equations would be extremely pleasing. In addition, gas particles actually flow with low friction outside the boundary area. Therefore, we will explore this possibility.

A *potential flow* is both inviscid and irrotational.
A flow has a *velocity potential* ϕ if a scalar function ϕ satisfies grad $\phi = \mathbf{u}$.
The level sets $\phi = $ const are called *equipotential lines*.

Potential flow is an idealization in which friction and vortex formation *within the flow area* can be neglected. The term has a historical origin and can cause confusion. Not every potential flow actually has a velocity potential.

Vortices can form at the boundaries of obstacles, for example at the wing tip of an airfoil, where the flow exits at different speeds at the top and bottom (see Fig. 4.11).

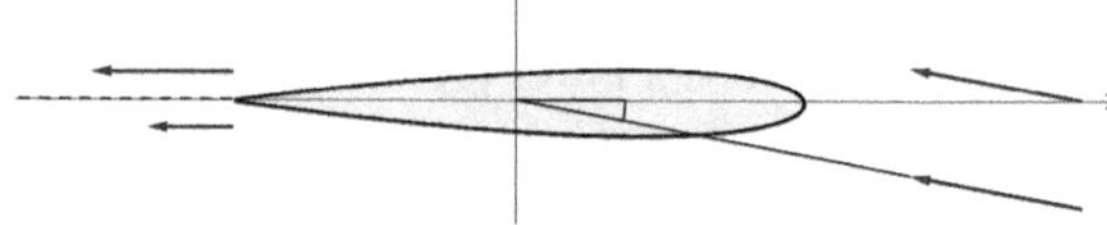

Fig. 4.11 A singularity is created at the tail of an airfoil (*left dashed line*).

Vortex transport is now possible at the *singularities* (discontinuities) of the velocity field, which are crucial for the generation of lift (see Sects. 5.6 and 5.7).

Problem 4.3 Prove that every vortex-free flow with constant density is a potential flow.

Problem 4.4 Prove that every potential flow $\mathbf{u}$ in a simply connected region has a velocity potential ϕ.

Let us examine the two-dimensional exterior domain of a fixed obstacle, for example an aerofoil profile. In this multiply connected region, we consider a stationary flow, which we additionally assume to be vortex-free and incompressible. Under these conditions, the velocity can be modelled as an analytic function, so that a rich mathematical theory is available. Therefore, we convert all vector fields into complex quantities. Section A.4 compiles important definitions and formulas.

The vectorial velocity $\mathbf{u} = \begin{pmatrix} u \\ v \end{pmatrix}$ is assigned the *complex velocity*

$$\mathcal{U} = u - \mathrm{i}v$$

Similarly, the force $\mathbf{F} = \begin{pmatrix} F_x \\ F_y \end{pmatrix}$ corresponds to the *complex force* $\mathcal{F} = F_x - \mathrm{i}F_y$.

Note that vectors are thus not converted into the immediate, but into conjugate complex quantities.

Lemma 4.1 *In a two-dimensional, stationary flow with a vortex-free, incompressible velocity field* $\mathbf{u} = \begin{pmatrix} u \\ v \end{pmatrix}$ *the complex velocity* $\mathcal{U} = u - \mathrm{i}v$ *is an analytic function.*

Proof Incompressibility means $\operatorname{div} \mathbf{u} = 0$, thus

$$\frac{\partial u}{\partial x} + \frac{\partial v}{\partial y} = 0$$

Considering (1.21), irrotationality rot $\mathbf{u} = 0$ implies that

$$\frac{\partial v}{\partial x} - \frac{\partial u}{\partial y} = 0$$

Thus, for $\mathcal{U} = u + \mathrm{i}(-v)$ we obtain the Cauchy-Riemann differential equations

$$\frac{\partial u}{\partial x} = -\frac{\partial v}{\partial y}, \qquad \frac{\partial v}{\partial x} = \frac{\partial u}{\partial y},$$

which means that this function is analytic. $\qquad\qquad\qquad\qquad\qquad\qquad\square$

Problem 4.5 Prove that every incompressible flow with a velocity potential is inviscid.

Problem 4.6 We consider a non-stationary flow of constant density with conservative force density $\mathbf{f}/\varrho = \mathrm{grad}\,\Phi$ and velocity potential ϕ. Prove the non-stationary Bernoulli equation

$$\frac{\partial \phi}{\partial t} + \frac{1}{2}\|\mathrm{grad}\,\phi\|^2 + \frac{p}{\varrho} + \Phi = c(t), \tag{4.8}$$

where $c(t)$ is a time-dependent constant.

4.3   The Kutta-Joukowski Theorem

We consider a flow around a fixed obstacle $D \subset \mathbb{C}$, whose boundary is denoted by $C = \partial D$ and forms a simple closed curve. For simplification, we assume that the origin of coordinates is in D.

The following theorem represents the first step in calculating the lift of airfoil profiles. Here, the force acting on an obstacle is calculated from the pressure, which is coupled to the velocity via Bernoulli's theorem.

Theorem 4.2 (Blasius) *We consider an obstacle with boundary C in a potential flow of constant density. Then, the force acting on the obstacle is given by*

$$\mathcal{F} = -\frac{\mathrm{i}\varrho}{2}\overline{\left[\int_C \mathcal{U}^2\,\mathrm{d}z\right]}$$

Proof We consider the infinitesimal increment $\mathrm{d}z = \mathrm{d}x + \mathrm{i}\,\mathrm{d}y$ along C and the corresponding increment $\mathrm{d}z/\mathrm{i}$ in the normal direction (see Exercise A.6). Since

$$\mathbf{F} = -\int_C p\mathbf{n}\,\mathrm{d}s,$$

the complex force is given by

$$\mathcal{F} = -\int_C p \, \frac{\mathrm{d}z}{\mathrm{i}} = \mathrm{i} \int_C p \, \mathrm{d}z$$

Because potential flows are vortex-free, we can apply Bernoulli's theorem (Theorem 3.2) to the exterior region $\mathbb{R}^2 \setminus D$ and obtain

$$\frac{p}{\varrho} + \frac{u^2 + v^2}{2} = K \qquad \text{with a constant } K$$

Adding a constant to p does not change the force integral. We can therefore set $K = 0$, i.e.

$$p = -\frac{\varrho(u^2 + v^2)}{2} \tag{4.9}$$

and

$$\mathcal{F} = -\frac{\mathrm{i}\varrho}{2} \int_C (u^2 + v^2) \, \mathrm{d}z$$

On the other hand, we have $\mathcal{U}^2 = (u - \mathrm{i}v)^2 = u^2 - v^2 - 2\mathrm{i}uv$. At the boundary C, the velocity $\mathbf{u}$ is parallel to the tangential vector $\begin{pmatrix} \mathrm{d}x \\ \mathrm{d}y \end{pmatrix}$ and consequently $u \, \mathrm{d}y = v \, \mathrm{d}x$. Thus,

$$\mathcal{U}^2 \, \mathrm{d}z = (u^2 - v^2 - 2\mathrm{i}uv)(\mathrm{d}x + \mathrm{i}\mathrm{d}y) = (u^2 + v^2)(\mathrm{d}x - \mathrm{i}\mathrm{d}y)$$

Since $u^2 + v^2$ is real, it holds

$$\overline{\mathcal{U}^2 \, \mathrm{d}z} = (u^2 + v^2)\overline{(\mathrm{d}x - \mathrm{i}\mathrm{d}y)} = (u^2 + v^2) \, \mathrm{d}z$$

By substitution into the above integral, we obtain the assertion. $\qquad\square$

The circulation, which is closely related to vorticity, will prove to be a crucial factor in calculating lift.

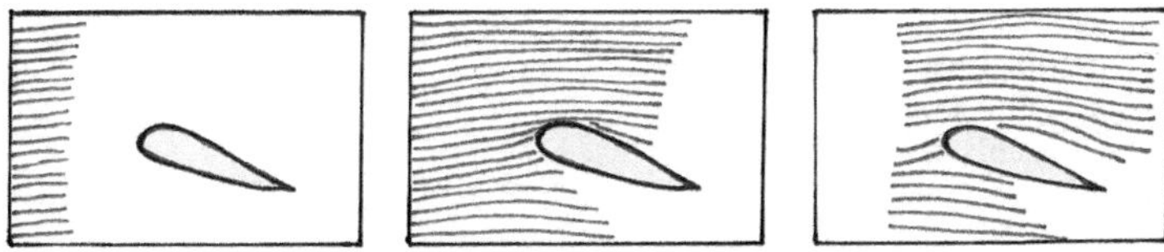

Fig. 4.12 Flow with smoke particles around an airfoil profile.

The *circulation* round a closed curve C is defined by

$$\Gamma = \oint_C \mathbf{u}\,\mathrm{d}\mathbf{x} \tag{4.10}$$

The index C is omitted when we integrate around the boundary of the obstacle.

The *inflow velocity* refers to the flow velocity at infinity.

$$\mathbf{u}_\infty = \begin{pmatrix} U \\ V \end{pmatrix} \qquad \text{or} \quad \mathcal{U}_\infty = U - iV \tag{4.11}$$

Problem 4.7 Figure 4.12 shows the flow around an airfoil at successive points in time. Estimate the sign of the circulation around the boundary.

By rewriting into complex functions, we obtain

$$\Gamma = \oint_C \mathbf{u}\,\mathrm{d}\mathbf{x} = \oint_C u\,\mathrm{d}x + v\,\mathrm{d}y = \oint_C (u - iv)\,(\mathrm{d}x + i\,\mathrm{d}y)$$

$$= \oint_C \mathcal{U}\,\mathrm{d}z \tag{4.12}$$

Let $\mathbf{n}$ denote the normal vector to $\mathbf{u}_\infty$. Now we can derive the lift formula.

Theorem 4.3 (Kutta-Joukowski) *In the exterior region around an obstacle with boundary C, we consider a potential flow of constant density with the inflow velocity $\mathbf{u}_\infty$. The force acting on the obstacle is given by*

$$\mathbf{F} = \varrho\Gamma\,\|\mathbf{u}_\infty\|\,\mathbf{n}, \tag{4.13}$$

where the circulation Γ is calculated around C.

If the obstacle has a shape that allows sufficient circulation around its boundary and potential flow in the exterior region, then it is suitable as an airfoil profile.

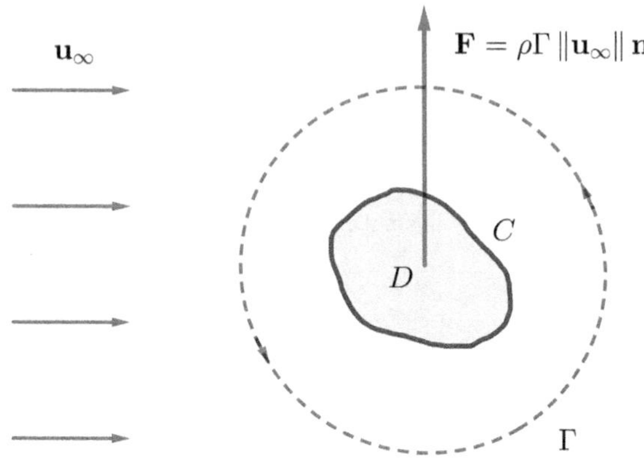

Fig. 4.13 In the case of horizontal inflow, the force acts in the vertical direction. If it is directed upwards, it is referred to as lift.

This sounds simple, but the practical implementation required decades of demanding experiments.

Proof By assumption, the origin of the coordinates is located in the obstacle D (Fig. 4.13). In the flow-through area, $\mathcal{U}$ is an analytic function and can be represented as a Laurent series (Theorem A.2). Because $\lim_{z \to \infty} \mathcal{U} = \mathcal{U}_\infty$, the series expansion must not contain any positive powers of z. Then for every circle with origin zero, which contains the boundary C

$$\mathcal{U} = a_0 + \frac{a_1}{z} + \frac{a_2}{z^2} + \frac{a_3}{z^3} + \dots, \quad \mathcal{U}^2 = a_0^2 + \frac{2a_0 a_1}{z} + \frac{2a_0 a_2 + a_1^2}{z^2} + \dots$$

By Problem A.8, it holds

$$\int_C \frac{dz}{z^k} = \begin{cases} 2\pi i & \text{if } k = 1 \\ 0 & \text{if } k \neq 1 \end{cases}$$

and we obtain

$$\int_C \mathcal{U}^2 \, dz = 4\pi i \, a_0 a_1$$

We calculate the remaining coefficients. Considering $\lim_{z \to \infty} \mathcal{U} = \mathcal{U}_\infty$ and (4.11), it follows that

$$a_0 = U - iV$$

Using Problem A.8 again, we find $\Gamma = \oint_C \mathcal{U} \, dz = 2\pi i \, a_1$ and thus

$$a_1 = \frac{\Gamma}{2\pi i}$$

Applying the Blasius theorem, we get

$$\mathcal{F} = -\frac{i\varrho}{2}\overline{\left[\int_C \mathcal{U}^2\,dz\right]} = -\frac{i\varrho}{2}\overline{(4\pi i\, a_0 a_1)} = \varrho\Gamma(V - iU)$$

Without loss of generality, we choose the coordinate system such that the flow is in the positive x-direction, i.e. $U = \|\mathbf{u}_\infty\|$, $V = 0$. Then

$$\mathcal{F} = -i\varrho\Gamma\,\|\mathbf{u}_\infty\|$$

Since the multiplication $-i$ causes a rotation by a right angle, the assertion follows by transforming the complex quantity back into vector form. $\square$

Whether the vertical force (4.13) is directed upwards or downwards depends on the sign of Γ and thus also on the geometry of the obstacle. In [2], an analogue to Theorem 4.3 was proven for the practically significant *premature flow separation (Helmholtz-Kirchhoff flow)*.

The concept of circulation is understood through the following remarks.

- If we were to use the more precise Navier-Stokes equations instead of the Euler equations, the circulation would be zero, as would the boundary velocity of the Blasius theorem. The circulation from (4.13), an "approximation" that is better than the exact value, requires interpretation. It can be recognized from the proof of the Blasius theorem and the underlying Euler equations:

  The boundary velocity of the Euler equations is not an approximation of the velocity to the boundary in the sense of a mathematical limit, but a projection of the flow behavior averaged in a surrounding layer at the boundary. The circulation (4.10) is another averaging and thus describes the near-boundary behaviour in a single numerical value.
- Using (4.13), we have traced the two-dimensional lift back to the presence of a circulation. The same will be found in Sect. 4.8 for the three-dimensional case, without anything being said about the causes of circulation. Its origin can be explained by friction in the boundary layer followed by vortex formation (see Sect. 5.6).

To simplify calculations, it may be useful to imagine an obstacle as being replaced by fluid that has a suitable vortex distribution. This is referred to as a *bound vortex*.

In contrast, we speak of *free* vortices, which can be explained without the aforementioned conceptual change of reality.

Bound vortices can be introduced at obstacles when the circulation Γ does not disappear, for example at airfoils. According to Stokes' theorem (Theorem 1.2), the virtual fluid fulfils the condition $\Gamma = \int_S (\operatorname{rot} \mathbf{u}) \cdot \mathbf{n}\, dS$, where S denotes the boundary of the obstacle.

4.4 Complex Potentials

At the beginning of the twentieth century, the lift of various profiles was calculated using complex potentials. Although this technique remained limited to special cases, it had the advantage over later numerical methods of providing insight into the nature of the processes. Here we explain the basics. First, we consider a two-dimensional potential flow from Sect. 4.2, where $\mathcal{U} = u - iv$ denotes its complex velocity field.

> If the analytic function $\mathcal{U} = u - iv$ in the domain $\Omega \subset \mathbb{C}$ has an analytic antiderivative $w = \phi + i\psi$, this is referred to as its *complex potential*.

Lemma 4.2 *A flow $\mathcal{U}$ with complex potential $w = \phi + i\psi$ has the following properties.*

(1) ψ is the stream function and ϕ is the velocity potential of $\mathbf{u}$.
(2) The streamlines $\psi = $ const are orthogonal to the equipotential lines $\phi = $ const.
(3) The flow is irrotational.

Proof

Ad. (1): For $z = x + iy$, $w(z) = \phi(x, y) + i\psi(x, y)$ it follows that

$$\frac{\partial \phi}{\partial x} + i\frac{\partial \psi}{\partial x} = \frac{\partial w}{\partial x} = \frac{dw}{dz}\frac{\partial z}{\partial x} = \frac{dw}{dz} = \mathcal{U} = u - iv,$$

hence $\frac{\partial \phi}{\partial x} = u$, $\frac{\partial \psi}{\partial x} = -v$. Using the Cauchy-Riemann differential equations for w, we get

$$\frac{\partial \phi}{\partial x} = \frac{\partial \psi}{\partial y}, \qquad \frac{\partial \phi}{\partial y} = -\frac{\partial \psi}{\partial x} \tag{4.14}$$

In summary, we have

$$u = \frac{\partial \phi}{\partial x} = \frac{\partial \psi}{\partial y}, \qquad v = \frac{\partial \phi}{\partial y} = -\frac{\partial \psi}{\partial x} \tag{4.15}$$

Thus, $\operatorname{grad} \phi = \mathbf{u}$, i.e. ϕ is a velocity potential. From Corollary 2.2 it follows that ψ is a stream function.

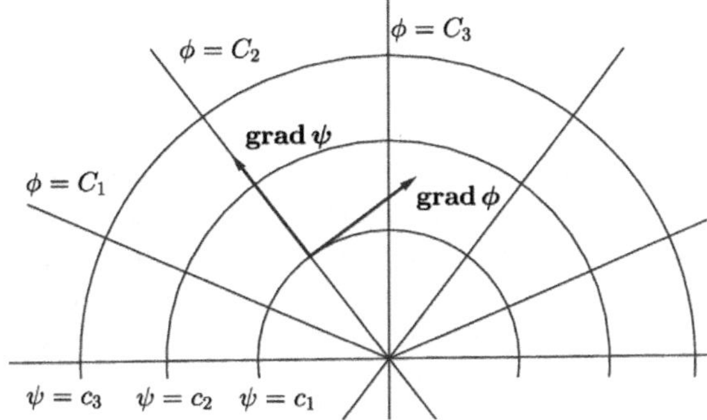

Fig. 4.14 If the flow $\mathcal{U}$ has a complex potential, then the streamlines and the equipotential lines form an orthogonal network.

Ad. (2):   The orthogonality (Fig. 4.14) follows from (4.14) by

$$\operatorname{grad}\phi \cdot \operatorname{grad}\psi = \frac{\partial\phi}{\partial x}\frac{\partial\psi}{\partial x} + \frac{\partial\phi}{\partial y}\frac{\partial\psi}{\partial y} = \frac{\partial\phi}{\partial x}\left(-\frac{\partial\phi}{\partial y}\right) + \frac{\partial\phi}{\partial y}\frac{\partial\phi}{\partial x} = 0$$

Ad. (3):   From

$$\frac{\partial u}{\partial y} = \frac{\partial}{\partial y}\frac{\partial\phi}{\partial x}, \qquad \frac{\partial v}{\partial x} = \frac{\partial}{\partial x}\frac{\partial\phi}{\partial y}$$

we get $\frac{\partial v}{\partial x} - \frac{\partial u}{\partial y} = 0$. The statement follows from (1.21).

$\square$

Problem 4.8 Show that the imaginary part of the complex potential remains constant on the boundary of the domain.

The flow of the potential

$$w(z) = \frac{i\Gamma}{2\pi}\ln(z - z_0), \qquad \Gamma = \text{const} \tag{4.16}$$

is referred to as a *potential vortex* at z_0.

Problem 4.9 Show that the potential vortex (4.16) describes a flow rotating around its centre z_0 (see Fig. 4.14).

Using (4.16) and Lemma 4.2, we can model vortices that are irrotational in $\mathbb{C} \setminus \{z_0\}$ and are therefore easier to handle. This approach opens up perspectives for interesting applications.

The method of complex potentials offers a practical alternative to the Navier-Stokes equations when friction is mainly confined to a narrow layer at the boundary. Of the Euler equations (3.30), (3.31), only (3.31) for *incompressibility* is retained, while the equation of motion (3.30) is replaced by the requirement for *irrotationality* rot $\mathbf{u} = 0$. Both conditions are met by the requirement of a complex potential w. Thus, the differential equations are replaced by the search for w, which even takes into account the circulation when including a suitable term. Owing to Problem 4.8, the boundary condition (3.32) is satisfied by the requirement that the imaginary part of w remains constant on the boundary. For unbounded domains, a constant velocity at infinity is additionally required, i.e. $\lim_{z\to\infty} \mathrm{d}w/\mathrm{d}z = \text{const}$.

Without (3.30) the potential model lacks an equation for the pressure p, which is needed for the force $\mathbf{F} = \int_C p \, \mathrm{d}\mathbf{x}$. The bottleneck can be circumvented with Bernoulli's theorem (Theorem (3.2)), which allows the pressure calculation from the velocity. Since Bernoulli's theorem was derived on the basis of (3.30), the method of complex potentials in the extended sense also includes Euler's equation of motion.

A conformal mapping $z = \mathcal{E}(\zeta)$ transforms potentials into each other:

$$\hat{w}(z) = w(\mathcal{E}^{-1}(z)) \tag{4.17}$$

If the treatment of $w(\zeta)$ is simpler than that of $\hat{w}(z)$, we start the calculations in the ζ- coordinates. The velocity in z-coordinates can then be calculated using

$$\frac{\mathrm{d}\hat{w}}{\mathrm{d}z} = \frac{\mathrm{d}w}{\mathrm{d}\zeta} \frac{\mathrm{d}\zeta}{\mathrm{d}z} \tag{4.18}$$

Problem 4.10 For an obstacle, two different stationary flows with the complex potentials w_1 and w_2 are known. Show for arbitrary constants a_1, a_2, that $a_1 w_1 + a_2 w_2$ also describes a flow around the obstacle.

4.5   Flow around a Circular Cylinder

The search for a complex potential is complicated by the requirement that its imaginary part must be constant on a given boundary C. The problem was first solved for a circle. From there, transformations into ellipses and other interesting profiles were found.

$K(C, a) \subset \mathbb{C}$ denotes a circle with radius a and centre C. It is described by $|\zeta - \zeta_c| = a$, where ζ_c are the centre coordinates.

As an obstacle in a two-dimensional flow, it is called a *circular cylinder*.

The following statement is also called the *circle theorem*.

Theorem 4.4 (Milne-Thomson) *In the complex plane, a flow is given by its complex potential $w(z) = f(z)$, where $|z| > a$ holds for all its singularities.*

After inserting a solid circular cylinder $|z| = a$, the changed flow can be described by the potential

$$w_a(z) = f(z) + \bar{f}\left(\frac{a^2}{z}\right)$$

where $\bar{f}$ denotes the complex conjugate of f (see Sect. A.4).

Proof It holds $|a^2/z| > a$ for all singularities of $\bar{f}(a^2/z)$. Thus, they lie inside the obstacle $|z| = a$ and do not influence the flow. On the boundary of the circular cylinder, $z\bar{z} = a^2$ and therefore

$$w_a(z) = f(z) + \bar{f}(\bar{z}) = f(z) + \overline{f(z)} \in \mathbb{R}$$

We therefore find that $\psi = 0$ on $|z| = a$, i.e. the fluid cannot penetrate the boundary.
$\square$

As will be shown later, lift calculations in the complex plane can be reduced to the flow around a circle. The following serves as a cornerstone.

Example 4.1 In search of a flow around a circular cylinder, we first consider a uniform flow with the velocity

$$\mathbf{u} = u_\infty \begin{pmatrix} \cos\alpha \\ -\sin\alpha \end{pmatrix},$$

which is inclined at an angle α to the x-axis. It has the complex velocity

$$\mathcal{U}_1 = u_\infty[\cos\alpha + i\sin\alpha] = u_\infty e^{i\alpha}$$

and the potential

$$w_1(z) = u_\infty e^{i\alpha} z$$

By inserting the obstacle $|z| = a$ we obtain for the changed flow from Theorem 4.4 the potential

$$w_2(z) = u_\infty \left[z e^{i\alpha} + \frac{a^2 e^{-i\alpha}}{z} \right] \tag{4.19}$$

According to the Problems 4.9, 4.10 we can add a potential vortex at $z = 0$, i.e.

$$w_3(z) = u_\infty \left[z e^{i\alpha} + \frac{a^2 e^{-i\alpha}}{z} \right] - \frac{i\Gamma}{2\pi} \ln z, \qquad \Gamma = \text{const}$$

represents a further potential for the given obstacle. We shift it from the origin to z_c. After renaming z, z_c to ζ, ζ_c we get

$$w(\zeta) = u_\infty \left[(\zeta - \zeta_c) e^{i\alpha} + \frac{a^2 e^{-i\alpha}}{\zeta - \zeta_c} \right] - \frac{i\Gamma}{2\pi} \ln(\zeta - \zeta_c), \quad |\zeta - \zeta_c| \geq a \tag{4.20}$$

Problem 4.11 Prove that (4.20) implies a constant velocity at a sufficient distance from the boundary $K(C, a)$.

$$\lim_{\zeta \to \infty} \mathcal{U}(\zeta) = u_\infty e^{i\alpha}$$

According to (4.13), our potential only allows lift if there is circulation around the profile.

Problem 4.12 Prove that Γ in (4.20) indicates the circulation around $K(C, a)$.

We want to show that the value of the circulation changes the velocity field qualitatively. The stagnation points provide an important clue. For simplicity we set $\alpha = 0$.

Lemma 4.3 *Consider the flow given by (4.20). Depending on the value*

$$\kappa = -\frac{\Gamma}{2\pi a u_\infty}, \tag{4.21}$$

the velocity field $\mathcal{U} = \frac{dw}{d\zeta}$ *has the following properties.*

(1)   Let $\kappa < 2$. *Further, let* β *be defined by*

$$\sin \beta = \frac{\kappa}{2} \tag{4.22}$$

Then, we find two stagnation points

$$\zeta = \zeta_c + a(-i \sin \beta \pm \cos \beta) \tag{4.23}$$

They are symmetrically to each other on the circle (Fig. 4.15).

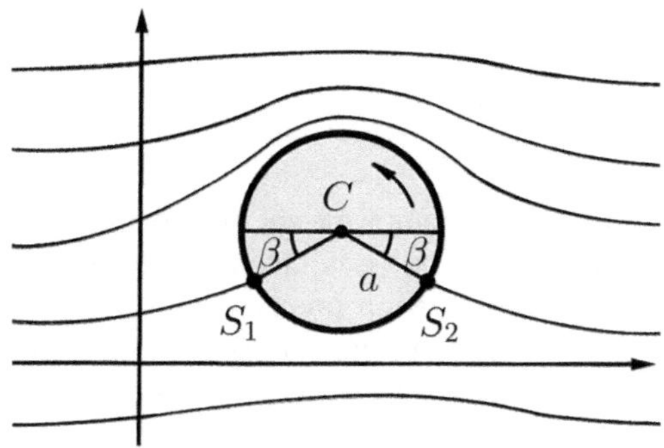

Fig. 4.15 Streamlines of (4.20) for $\kappa < 2$. Owing to its stagnation points, the model can be used to calculate lift.

(2)   Let $\kappa = 2$. Then, we get a stagnation point on the circle

$$\zeta = \zeta_c - a\mathrm{i} \tag{4.24}$$

(3)   Let $\kappa > 2$. Further, let β be defined by

$$\cosh \beta = \frac{\kappa}{2}, \quad \beta \neq 0 \tag{4.25}$$

Consider the points

$$\zeta = \zeta_c + a\mathrm{i}(-\cosh \beta \pm \sinh \beta) = \zeta_c - a\mathrm{i}e^{\pm\beta} \tag{4.26}$$

Then, one of them lies in the flow area and represents a stagnation point, while the other lies outside the flow area.

Proof Let us calculate the stagnation points of (4.20), (4.21). From $\mathcal{U} = 0$ follows

$$0 = u_\infty \left[1 - \frac{a^2}{(\zeta - \zeta_c)^2} + \frac{\mathrm{i}a\kappa}{\zeta - \zeta_c} \right]$$

$$\zeta = \zeta_c + a\left[-\frac{\mathrm{i}\kappa}{2} \pm \sqrt{1 - \frac{\kappa^2}{4}} \right] \tag{4.27}$$

Considering the discriminant, we distinguish between the following cases.

$\kappa/2 < 1$.   By substitution $\sin \beta = \kappa/2$ follows (4.23), see Fig. 4.15.
$\kappa/2 = 1$.   From (4.27) follows (4.24).
$\kappa/2 > 1$.   By substitution $\cosh \beta = \kappa/2$ follows (4.26). For the solutions ζ_1 and ζ_2, we have

$$|\zeta_1 - \zeta_c| = ae^{\beta} \quad \text{or} \quad |\zeta_2 - \zeta_c| = ae^{-\beta}$$

If $e^\beta > 1$, then $e^{-\beta} < 1$ and ζ_1 is the only stagnation point. In the opposite case $e^{-\beta} > 1$, the only stagnation point is at ζ_2.

$\square$

In the following we want to convert our circular profile (4.20) into a wing profile. Their flow behavior is characterized by two special features, see Fig. 4.8. First, there is a stagnation point below the rounded nose that needs to be transformed into a stagnation point of the circular profile. Secondly, the boundary at the tail is not smooth and the velocity becomes finite there only if the transformed point on the circular profile is a stagnation point. Therefore, in Lemma 4.3 only case (1) with two stagnation points is suitable for a transformation into a wing flow. From (4.20),(4.21) and (4.22) we obtain the approach

$$w(\zeta) = u_\infty\left[(\zeta - \zeta_c) + \frac{a^2}{\zeta - \zeta_c} + \mathrm{i}\, 2a \sin\beta \ln(\zeta - \zeta_c)\right] \qquad (4.28)$$

$$\sin\beta = -\frac{\Gamma}{4\pi a u_\infty}, \qquad |\zeta - \zeta_c| \geq a$$

Example 4.2 The *Magnus effect* describes the lift that a rotating object experiences in a flow. The rotation of a solid in an air stream changes the speed of the incoming air particles, so that they are accelerated on one side of the object and decelerated on the opposite side. According to (4.13), a non-vanishing circulation occurs at the boundary of the object, causing a force. (In example 4.1 we have introduced a potential that allows corresponding calculations.)

This principle is used in the *Flettner* or *rotor ship*, where motor-driven rotating towers support propulsion in a crosswind (see Figs. 4.16 and 4.17). If the wind blows in the opposite direction, the direction of rotation of the towers must be changed. Flettner ships are characterised by low energy consumption while maintaining seaworthiness.

Fig. 4.16 The Buckau was a converted schooner and the first rotor ship. It was designed in 1924 by Anton Flettner in collaboration with Ludwig Prandtl.

Fig. 4.17 Top view of a Flettner ship. Its propulsion is based on the different flow velocities around the rotor, which generate a non-vanishing circulation.

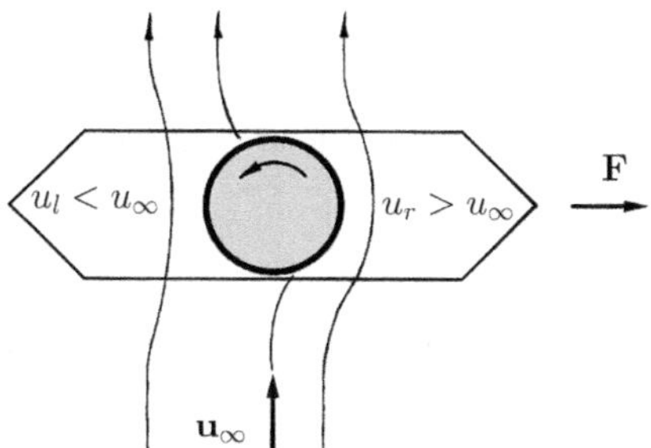

The savings compared to a purely conventional drive are between 5 and 20%. Currently, Flettner ships are used as container ships or ferries. One of the largest is the tanker Maersk Pelican with a deadweight (DWT) of 109,647 t.

4.6   Flow around an Elliptical Cylinder

The idea of the following model has already been illustrated in Fig. 4.1. It can be found in various situations.

- In aerodynamics, a flattened ellipse can be used as a wing model. By calculating the position of the stagnation points at take off, the formation of the starting vortex in Sect. 5.6 becomes more clearly understandable.
- From navigation it is known that the helmsman must correct the course of his ship even when it is sailing in the direction of the current. We explain why a rudderless ship inevitably positions itself perpendicular to the current.

To describe a force acting on a rigid body, we need not only its effect on the center of gravity but also its torque. For this purpose we supplement Blasius' theorem (Theorem 4.2). As in Sect. 4.4, we use two-dimensional flows with a complex potential w and the complex velocity $\mathrm{d}w/\mathrm{d}z = \mathcal{U} = u - iv$.

> We denote the *torque* on a profile by the vector **M**, which is oriented in the z-direction.
>
> Its z-component M indicates both the magnitude and the direction ($M > 0$ for the mathematically positive direction of rotation or orientation in the positive z-direction).

Problem 4.13 We consider the boundary point $Q = \begin{pmatrix} x \\ y \end{pmatrix}$ of a flow-around profile. Prove for the torque around the origin that Q contributes

$$dM = p(x\,dx + y\,dy) \tag{4.29}$$

Problem 4.14 Prove for an analytic function w

$$\frac{d\bar{w}}{d\bar{z}} = \overline{\left(\frac{dw}{dz}\right)} \tag{4.30}$$

Problem 4.15 Let $w = \phi + i\psi$ be a complex potential in a domain with boundary C. Show that

$$d\bar{w} = dw \qquad \text{on } C \tag{4.31}$$

Theorem 4.5 (Blasius) *Consider a potential flow of constant density around an obstacle with boundary C. Let w be its complex potential. Then the following torque is applied to the obstacle around the origin*

$$M = -\frac{\varrho}{2}\,\mathrm{Re}\left[\int_C z\left(\frac{dw}{dz}\right)^2 dz\right]$$

Proof By (4.29), we have

$$dM = p(x\,dx + y\,dy) = \mathrm{Re}\,[pz\,d\bar{z}] \tag{4.32}$$

As in the proof of Theorem 4.2, we use Bernoulli's theorem in the form (4.9)

$$p = -\frac{\varrho(u^2 + v^2)}{2} \tag{4.33}$$

Taking into account $dw/dz = u - iv$ and (4.30), we find that

$$u^2 + v^2 = (u - iv)(u + iv) = \frac{dw}{dz}\overline{\left(\frac{dw}{dz}\right)} = \frac{dw}{dz}\frac{d\bar{w}}{d\bar{z}} \tag{4.34}$$

Using (4.33), (4.34), we obtain

$$pz\,d\bar{z} = -\frac{\varrho}{2}\frac{dw}{dz}\frac{d\bar{w}}{d\bar{z}}z\,d\bar{z} = -\frac{\varrho}{2}\frac{dw}{dz}z\,d\bar{w}$$

Owing to (4.31), on the boundary C,

$$\frac{dw}{dz}\,d\bar{w} = \frac{dw}{dz}\,dw = \frac{dw}{dz}\frac{dw}{dz}\,dz$$

$$pz\,d\bar{z} = -\frac{\varrho}{2}\left(\frac{dw}{dz}\right)^2 z\,dz$$

and thus, after integrating (4.32), we get

$$M = -\frac{\varrho}{2}\,\mathrm{Re}\left[\int_C z\left(\frac{dw}{dz}\right)^2 dz\right]$$

$\square$

To find the complex potential, we proceed as described in Sect. 4.4. If we model the plan view of a ship's hull as an ellipse, it is recommended to use suitable coordinates.

The conformal mapping $z = \mathcal{E}(\zeta)$ is defined by

$$z = c\cosh\zeta \qquad \text{with } z = x + iy, \quad \zeta = \xi + i\eta \tag{4.35}$$

Here, (ξ, η) are referred to as *elliptic coordinates*.

Problem 4.16 Show that

$$\frac{x^2}{c^2\cosh^2\xi} + \frac{y^2}{c^2\sinh^2\xi} = 1 \tag{4.36}$$

$$\frac{x^2}{c^2\cos^2\eta} - \frac{y^2}{c^2\sin^2\eta} = 1 \tag{4.37}$$

For fixed ξ and η, the Eqs. (4.36) and (4.37) describe an ellipse and a hyperbola, respectively.

By choosing the parameter value $\xi_0 > 0$ for the boundary, we define an exterior domain.

In the ζ-plane, we consider the region $\xi \geq \xi_0$. For fixed α, we define the *potential*

$$w(\zeta) = cA \cosh(\zeta - \zeta_0), \qquad \zeta_0 = \xi_0 + i\alpha$$

and transform it into $\hat{w}(z) = w(\mathcal{E}^{-1}(z))$ using (4.17).

The interpretation of the parameters α, A is given in Problem 4.18. To calculate the force and the moment, we derive the potential.

Lemma 4.4 *It holds that*

$$\frac{d\hat{w}}{dz} = A\left(e^{-\zeta_0} - \frac{c^2 \sinh \zeta_0}{2z^2} + \dots\right) \tag{4.38}$$

$$\left(\frac{d\hat{w}}{dz}\right)^2 = A^2\left(e^{-2\zeta_0} - \frac{c^2 e^{-\zeta_0} \sinh \zeta_0}{z^2} + \dots\right), \tag{4.39}$$

where the dots stand for terms containing powers $\frac{1}{z^n}$, $n \geq 3$.

Proof Since

$$\frac{dw}{d\zeta} = cA \sinh(\zeta - \zeta_0), \qquad \frac{dz}{d\zeta} = c \sinh \zeta, \tag{4.40}$$

the chain rule (4.18) yields

$$\frac{d\hat{w}}{dz} = \frac{dw}{d\zeta}\frac{d\zeta}{dz} = \frac{dw}{d\zeta}\bigg/\frac{dz}{d\zeta} = \frac{A \sinh(\zeta - \zeta_0)}{\sinh \zeta} \tag{4.41}$$

Using the addition theorem for $\sinh(\zeta - \zeta_0)$, we get

$$\begin{aligned}
\frac{d\hat{w}}{dz} &= A\frac{\sinh \zeta \cosh \zeta_0 - \sinh \zeta_0 \cosh \zeta}{\sinh \zeta} \\
&= A\left[\cosh \zeta_0 - \sinh \zeta_0 \frac{\cosh \zeta}{\sinh \zeta}\right] \tag{4.42} \\
&= A\left[e^{-\zeta_0} + \sinh \zeta_0\left(1 - \frac{\cosh \zeta}{\sinh \zeta}\right)\right] \tag{4.43}
\end{aligned}$$

We simplify the last term. From (4.35) we obtain

$$\sqrt{z^2 - c^2} = c \sinh \zeta, \qquad \frac{\cosh \zeta}{\sinh \zeta} = \frac{z}{\sqrt{z^2 - c^2}}$$

Using the series expansion

$$\frac{1}{\sqrt{1-x}} = 1 - \frac{1}{2}x + \frac{1\cdot 3}{2\cdot 4}x^2 + \ldots \qquad \text{for } |x| < 1$$

we obtain

$$\frac{z}{\sqrt{z^2 - c^2}} = \frac{1}{\sqrt{1 - \frac{c^2}{z^2}}} = 1 - \frac{1}{2}\frac{c^2}{z^2} + \ldots$$

and thus

$$1 - \frac{\cosh\zeta}{\sinh\zeta} = \frac{1}{2}\frac{c^2}{z^2} - \ldots \tag{4.44}$$

From (4.43), (4.44), we find that

$$\frac{d\hat{w}}{dz} = A\left[e^{-\zeta_0} + \sinh\zeta_0\left(\frac{1}{2}\frac{c^2}{z^2} - \ldots\right)\right]$$

After squaring we get (4.39). $\qquad\qquad\qquad\square$

According to Sect. 4.4, we require that the flow remains uniform at infinity and does not penetrate the boundary. For this purpose, we check whether w satisfies the condition from Problem 4.8.

Problem 4.17 Show that the imaginary part of w remains constant for $\xi = \xi_0$.

Problem 4.18 Prove that $\hat{w}$ describes a uniform flow at infinity and show for the inflow velocity

$$\mathbf{u}_\infty = Ae^{-\xi_0}\begin{pmatrix} \cos\alpha \\ \sin\alpha \end{pmatrix} \tag{4.45}$$

$$u_\infty^2 = A^2 e^{-2\xi_0^2} \tag{4.46}$$

Thus, we have found an interpretation of the parameters α and A. From (4.45) it is clear that the inflow velocity $\mathbf{u}_\infty$ forms the angle α with the main axis of the obstacle.

Stagnation points provide important information about flow behavior and pressure distribution.

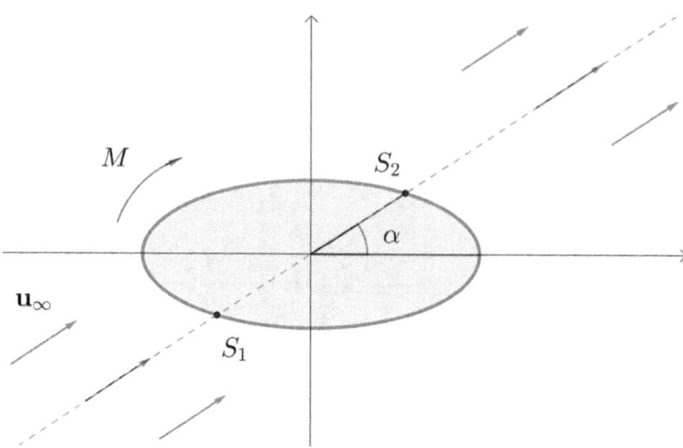

Fig. 4.18 Uniform flow around an elliptical obstacle with stagnation points S_1 and S_2, whereby the stagnation pressure measured at S_1 leads to the formation of a torque.

In a flow field around an obstacle, the *stagnation pressure* represents the highest pressure value and can be measured at one of the stagnation points.

Problem 4.19 Show that the stagnation points of $\hat{w}$ are at

$$\eta_1 = \alpha + \pi, \qquad \eta_2 = \alpha$$

When comparing Figs. 4.18 and 5.10, we notice that the positions of their stagnation points coincide. Thus, the flow around the ellipse describes an airfoil profile during takeoff, when the inflow velocity is still low.

Using Blasius's theorems we can calculate the resulting force and torque.

Theorem 4.6 *Consider the potential flow (4.17). Then the obstacle experiences neither lift nor drag, but the torque*

$$M = -\pi \varrho c^2 u_\infty^2 \sin \alpha \cos \alpha \tag{4.47}$$

Proof To calculate the forces, we apply Theorem 4.2. By $\mathcal{U} = \mathrm{d}\hat{w}/\mathrm{d}z$ and (4.39), we obtain

$$\mathcal{F} = -\frac{\mathrm{i}\varrho A^2}{2} \overline{\left[\int_C \left(\mathrm{e}^{-2\zeta_0} - \frac{c^2 \mathrm{e}^{-\zeta_0} \sinh \zeta_0}{z^2} + \ldots \right) \mathrm{d}z \right]}$$

$$= -\frac{\mathrm{i}\varrho A^2}{2} \overline{\left[\mathrm{e}^{-2\zeta_0} \int_C \mathrm{d}z \right]} \qquad \text{by Problem A.8}$$

$$= 0$$

For the torque, we get from Theorem 4.5 and (4.39)

$$
\begin{aligned}
M &= -\frac{\varrho}{2}\,\mathrm{Re}\left[\int_C zA^2\Big(\mathrm{e}^{-2\zeta_0} - \frac{c^2\mathrm{e}^{-\zeta_0}\sinh\zeta_0}{z^2} + \ldots\Big)\,\mathrm{d}z\right] \\
&= \frac{\varrho}{2}\,\mathrm{Re}\left[A^2 c^2 \mathrm{e}^{-\zeta_0}\sinh\zeta_0 \int_C \frac{\mathrm{d}z}{z}\right] \qquad \text{using Problem A.8 and Theorem A.1} \\
&= \frac{\varrho}{2}\,\mathrm{Re}\left[A^2 c^2 \mathrm{e}^{-\zeta_0}\sinh\zeta_0\, 2\pi i\right] \qquad \text{by Problem A.8} \\
&= -\varrho c^2 A^2 \pi\, \mathrm{e}^{-2\xi_0}\sin\alpha\,\cos\alpha
\end{aligned}
$$

After replacing (4.46), we obtain the assertion. $\qquad\qquad\qquad\qquad\square$

The unrealistic lack of lift and drag was to be expected since we were working with a complex potential without a circulation term. To interpret the torque, we first need the equilibria.

Problem 4.20 Show that the torque vanishes for $\alpha = 0$ (longitudinal position) and $\alpha = \pi/2$ (transverse position), where the first position is unstable and the second stable.

Although no torque occurs in the longitudinal position, it arises at the slightest course deviation. Owing to the sign in (4.47), the rotation continues until the ship's hull is perpendicular to the current. The effect can be seen in the flow patterns, as the stagnation pressure at S_1 causes a rotation to the right, see Fig. 4.18.

4.7 Lift Calculations on Profiles

Of the flow models at the beginning of Chap. 4, only the Kutta-Joukowski flow (Fig. 4.3) is suitable for describing stable flight behaviour. To calculate the lift, we again use the method from Sect. 4.4 (see Fig. 4.19).

Step 1: We look for a conformal mapping between the exterior regions of a circle and the profile to be examined. The circle points can thus be considered as coordinates of the profile.

Step 2: We choose a suitable complex potential in the exterior region of a circle and map it conformally into a complex potential for the flow around the wing.

Step 3: By deriving the potential, we obtain the velocity, from which the lift can be determined using the Blasius theorem. The calculation is performed by back-transformation into circular coordinates.

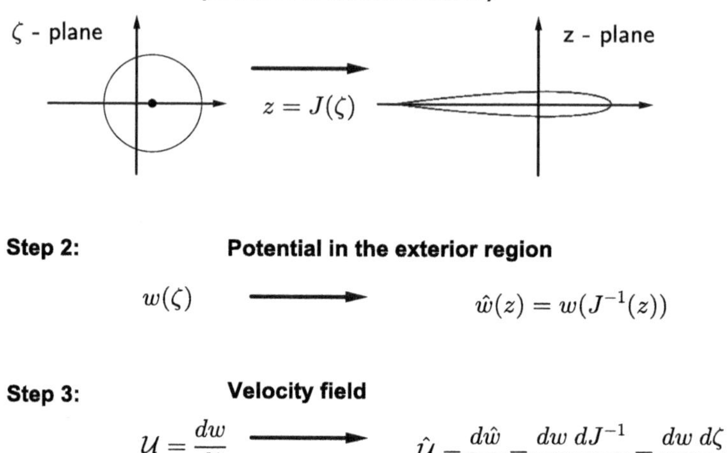

Fig. 4.19 calculation of the velocity field.

Regarding Step 1: Mapping a Circle onto a Wing Profile

The function

$$z = \zeta + \frac{1}{\zeta} \tag{4.48}$$

is called *Joukowski transformation.*

We want to prove that it transforms a suitable circle into a wing profile (Fig. 4.20).

Problem 4.21 Prove that the Joukowski transformation maps

(a) the circle $|\zeta| = 1$ onto the interval $[-2; 2]$,
(b) both the exterior region $|\zeta| > 1$ and the interior of the circle $|\zeta| < 1$ bijectively
     onto $\mathbb{C} \setminus [-2; 2]$.

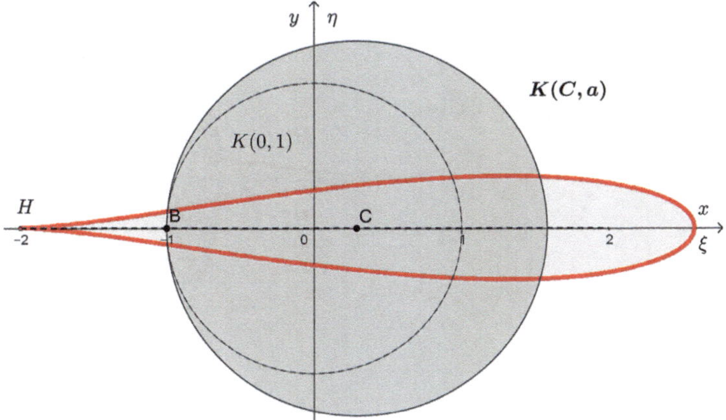

Fig. 4.20 Overlay of ζ- and z-plane. The Joukowski transformation maps the exterior region of the circle $K(C,a)$ bijectively onto the exterior region of a wing.

Fig. 4.21 The mapping (4.49) can be constructed by reflection on the unit circle and subsequent transition to the complex conjugate number.

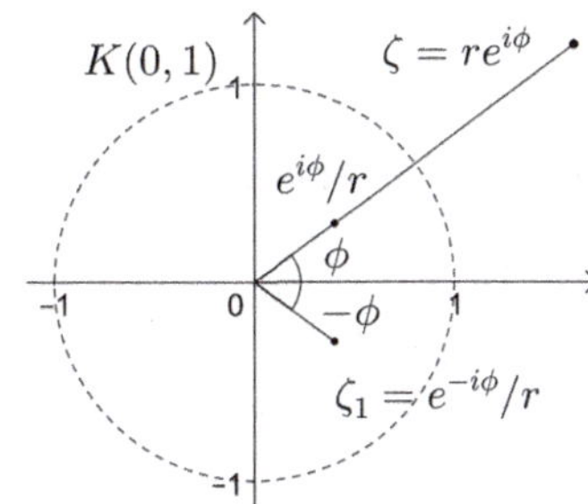

For a better understanding of the Joukowski transformation, we decompose it into

$$\zeta_1 = \frac{1}{\zeta} \tag{4.49}$$

$$z = \zeta + \zeta_1 \tag{4.50}$$

For $\zeta = re^{i\phi}$ we get $\zeta_1 = \frac{1}{r}e^{-i\phi}$. Thus, (4.49) is the composition of the reflection on the unit circle and the reflection on the abscissa, see Fig. 4.21. If the value ζ is passed through a suitable circle $K(C,a)$, we obtain from $\zeta + 1/\zeta$ the wing profile shown in Fig. 4.20.

Problem 4.22 We consider the circle $K(C,a)$ which contains the origin. Prove that (4.49) transforms this circle into a circle with centre $-\bar{\zeta}_0/\eta$ and radius a/η. ($\eta = a^2 - \zeta_0\bar{\zeta}_0$)

Problem 4.23 Prove that the Joukowki transformation is conformal in the exterior region $|\zeta| > 1$.

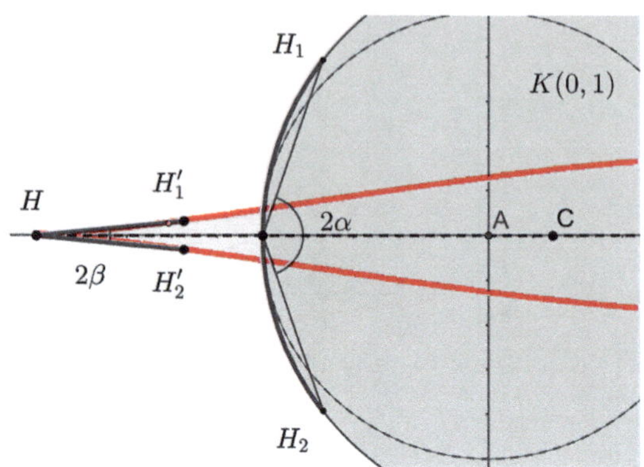

Fig. 4.22 Overlay of ζ- and z-plane. While the arc from H_1 to H_2 is traversed on $K(C, a)$, the image point of the Joukowski transformation moves on the trailing edge from H_1' via H to H_2'.

Problem 4.24 Prove the formula

$$\frac{z+2}{z-2} = \left(\frac{\zeta+1}{\zeta-1}\right)^2 \qquad \text{for } \zeta \neq 1 \tag{4.51}$$

We consider a circle $K(C, a)$ whose center is on the positive x-axis and whose left boundary point is at $\zeta = -1$ (Figs. 4.20 and 4.22). Since the points of the circle, with the exception of $\zeta = -1$, lie in the outer region $|\zeta| > 1$, their image under the Joukowski transformation is a smooth curve $z \neq -2$. The sharp angle at $z = -2$ can be explained by examining the transformation at $\zeta = -1$. In sufficiently small environments, we assume

$$\zeta = -1 + re^{i\alpha} \qquad \text{for the coordinates of } H_1 \in K(C, a)$$

$$z = -2 + Re^{i\beta} \qquad \text{for the coordinates of the image point } H_1'$$

Here, $r \ll 1$ and $R \ll 2$. By substitution into (4.51), we obtain approximately

$$\frac{Re^{i\beta}}{-4} = \left(\frac{re^{i\alpha}}{-2}\right)^2$$

$$Re^{i(\beta+\pi)} = r^2 e^{2\alpha i} \qquad \text{using } -1 = e^{i\pi}$$

Thus, $\beta + \pi = 2\alpha$. From $2\alpha \approx \pi$, it follows that $\beta \approx 0$. Therefore, our wing profile has a sharp angle at point H.

The following statement illustrates the effect of the Joukowski transformation in the exterior region of the unit circle.

Theorem 4.7 *The Joukowski transformation (4.48) maps concentric circles onto ellipses and rays onto hyperbolas (Figs. 4.23 and 4.24).*

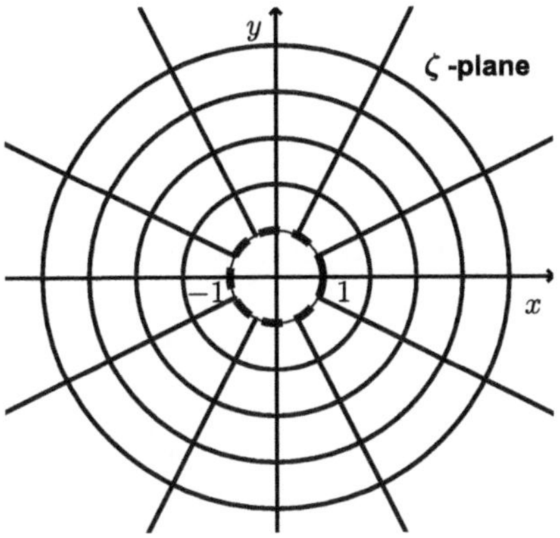

Fig. 4.23 Exterior region $|\zeta| > 1$ with radial and concentric grid lines.

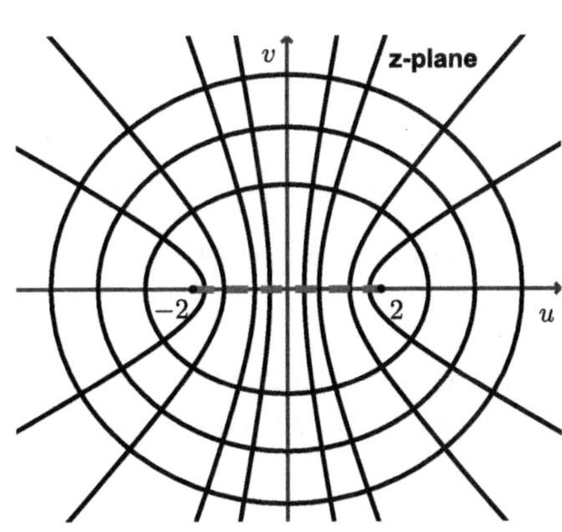

Fig. 4.24 The Joukowski transformation maps circ0les into ellipses and rays into hyperbolas.

Proof Under the assumption that $\zeta = r\mathrm{e}^{\mathrm{i}\phi}$, it follows from (4.48) that

$$z = r\mathrm{e}^{\mathrm{i}\phi} + \frac{1}{r}\mathrm{e}^{-\mathrm{i}\phi} = \left(r + \frac{1}{r}\right)\cos\phi + \mathrm{i}\left(r - \frac{1}{r}\right)\sin\phi$$

For $z = u + \mathrm{i}v$, we obtain

$$u = \left(r + \frac{1}{r}\right)\cos\phi, \qquad v = \left(r - \frac{1}{r}\right)\sin\phi \tag{4.52}$$

For a concentric circle in the ζ-plane, we have $r = \mathrm{const}$. Using $1 = \cos^2\phi + \sin^2\phi$, we derive from

$$\cos\phi = \frac{u}{\left(r + \frac{1}{r}\right)}, \qquad \sin\phi = \frac{u}{\left(r - \frac{1}{r}\right)},$$

the equation of the ellipse in the z-plane

$$1 = \frac{u^2}{(r + \frac{1}{r})^2} + \frac{v^2}{(r - \frac{1}{r})^2}$$

For a ray in the ζ-plane, we find that $\phi = $ const. From (4.52), we obtain

$$\frac{u^2}{\cos^2 \phi} = \left(r + \frac{1}{r}\right)^2 = r^2 + 2 + \frac{1}{r^2}$$

$$\frac{v^2}{\sin^2 \phi} = \left(r - \frac{1}{r}\right)^2 = r^2 - 2 + \frac{1}{r^2}$$

and by subtraction, the equation of the hyperbola in the z-plane

$$\frac{u^2}{\cos^2 \phi} - \frac{v^2}{\sin^2 \phi} = 4$$

The ellipses and hyperbolas have the focal points ± 2. $\square$

Regarding Step 2: Specification of a Potential Past the Circle and Transformation into a Potential for the Airfoil

We define a potential in the exterior area of the unit circle, which is to be mapped to the area outside the wing using the Joukowski transformation. With the exception of the wing tip, the mapping should be conformal in the exterior area of the wing.

The potential (4.28) contains the parameter β, which determines the circulation. The inflow is horizontal. For manoeuvrability, lift must be guaranteed even when the wing is tilted. Therefore, we introduce an angle of inclination α between the direction of inflow and the wing axis. For the calculations we rotate our coordinate system. (Figs. 4.25 and 4.26).

Fig. 4.25 In reality, the wing profile is tilted and the inflow is horizontal. The force acts perpendicularly.

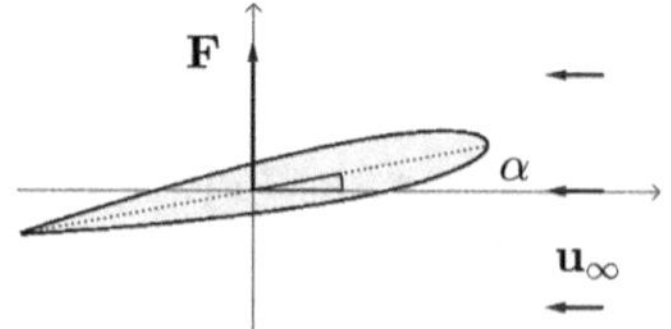

Fig. 4.26 For calculation, the coordinate system is rotated by α, so that the profile comes into a horizontal position (cf. Fig. 4.20).

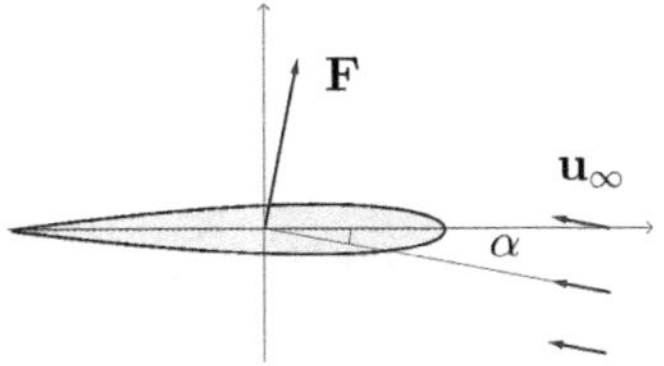

We consider a circle $K(C, a)$ with left boundary point at $\zeta = -1$ and center on the positive x-axis, so that $C = (\zeta_c, 0)$ for $\zeta_c > 0$ (Fig. 4.20). The function

$$w(\zeta) = u_\infty\left[(\zeta - \zeta_c)e^{i\alpha} + \frac{a^2 e^{-i\alpha}}{\zeta - \zeta_c} + i\,2a \sin\beta \ln(\zeta - \zeta_c)\right], \quad |\zeta - \zeta_c| \geq a$$

(4.53)

is referred to as the *circular potential* for $K(C, a)$.

Problem 4.25 Prove that the restriction of the Joukowski transformation to the exterior region $|\zeta - \zeta_c| \geq a$ has a unique inverse $\zeta(z)$.

This justifies the following definition.

We call the function

$$\hat{w}(z) = w(\zeta(z)) \tag{4.54}$$

the *potential of the wing profile*.
Here, $\zeta(z)$ denotes the inverse of the Joukowski transformation when restricted to the exterior region $|\zeta - \zeta_c| \geq 1$.

Regarding Step 3: Properties of the Associated Velocity Fields

We call the derivative

$$\mathcal{U}(\zeta) = \frac{dw}{d\zeta}, \quad |\zeta - \zeta_c| \geq a \tag{4.55}$$

the *velocity field past the circle.*
The derivative

$$\hat{\mathcal{U}}(z) = \frac{d\hat{w}}{dz} \tag{4.56}$$

is said to be the *velocity field past the wing profile.*

We want to show that $\hat{\mathcal{U}}$ is indeed a realistic velocity. From (4.53), (4.55), we find that

$$\mathcal{U}(\zeta) = u_\infty \left[e^{i\alpha} - \frac{a^2 e^{-i\alpha}}{(\zeta - \zeta_c)^2} + \frac{i\,2a \sin\beta}{\zeta - \zeta_c} \right] \tag{4.57}$$

By (4.54), (4.56), it follows that

$$\hat{\mathcal{U}}(z) = \frac{d}{dz} w(\zeta(z)) = \frac{dw}{d\zeta} \frac{d\zeta}{dz} = \mathcal{U}(\zeta) \frac{d\zeta}{dz} \tag{4.58}$$

Further, we obtain from (4.48)

$$\frac{d\zeta}{dz} = \frac{1}{dz/d\zeta} = \frac{1}{1 - \frac{1}{\zeta^2}}$$

In summary, it is

$$\hat{\mathcal{U}}(z) = \mathcal{U}(\zeta) \frac{1}{1 - \frac{1}{\zeta^2}} \tag{4.59}$$

$$= \frac{\mathcal{U}(\zeta)\,\zeta^2}{(\zeta - 1)(\zeta + 1)} \tag{4.60}$$

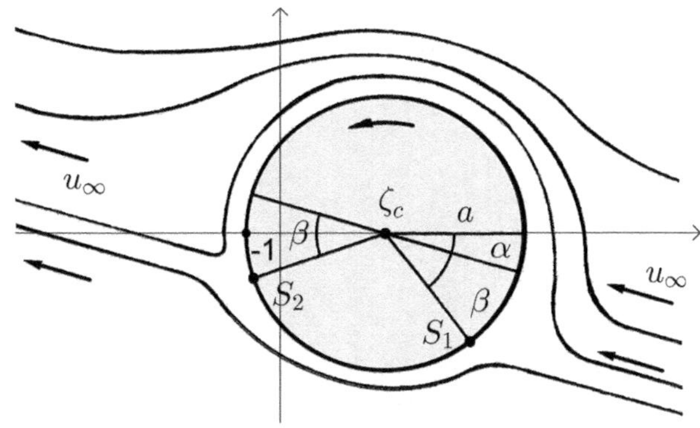

Fig. 4.27 Streamlines for the circular potential (4.53). The points S_1 and S_2 are stagnation points.

Problem 4.26 Prove that the velocity $\hat{\mathcal{U}}$ is constant at a sufficient distance from the wing, i.e.

$$\lim_{z \to \infty} \hat{\mathcal{U}}(z) = u_\infty e^{i\alpha} \tag{4.61}$$

According to the Kutta-Joukowski theorem (Theorem 4.3), lift requires a non-vanishing circulation around the wing.

Problem 4.27 Prove the formula $\Gamma = -4a\pi u_\infty \sin\beta$ for the circulation around the airfoil profile.

In the model (4.53), (4.54) we have not yet made any statement about the parameter β. It can be determined by the condition that the velocity $\hat{\mathcal{U}}$ must remain finite in the entire flow field. The problem is the wing tip $\zeta = -1$, where the denominator of (4.60) vanishes. The only way out is that there is a stagnation point for $\mathcal{U}$.

The requirement $\mathcal{U}(-1) = 0$ is called the *Joukowski condition*.
From Fig. 4.27 it can be seen that this condition is fulfilled for $\beta = \alpha$.

We demonstrate that the Joukowski condition ensures a finite velocity $\hat{\mathcal{U}}$ when the flow detaches at the tail.

Fig. 4.28 The flow leaves the wing tip in the direction of the profile axis at a finite speed.

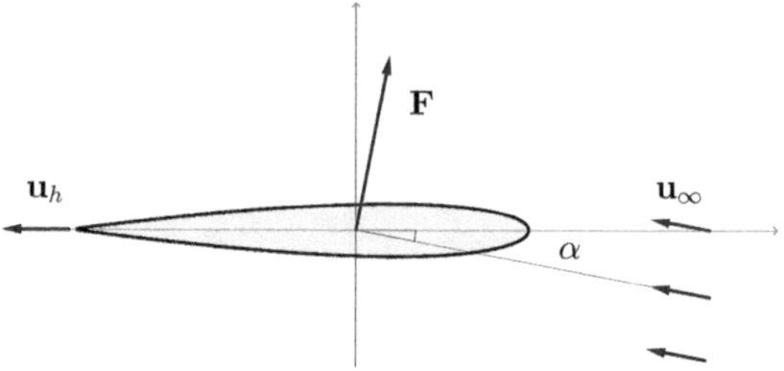

Theorem 4.8 *For $\beta = \alpha$, the flow leaves the wing tip $z = -2$ in the direction of the profile axis (Fig. 4.28) at the speed*

$$u_h = \frac{u_\infty \cos \alpha}{1 + \zeta_c} \tag{4.62}$$

Recall that $\zeta_c > 0$ denotes the x-coordinate of the circle centre C (Fig. 4.20) and α the angle of inclination between the inflow and the wing axis.

Proof Using $2\mathrm{i} \sin \beta = \mathrm{e}^{\mathrm{i}\beta} - \mathrm{e}^{-\mathrm{i}\beta}$, it follows from (4.57) that

$$\mathcal{U}(\zeta) = u_\infty \left[\mathrm{e}^{\mathrm{i}\alpha} - \frac{a^2 \mathrm{e}^{-\mathrm{i}\alpha}}{(\zeta - \zeta_c)^2} + \frac{a(\mathrm{e}^{\mathrm{i}\beta} - \mathrm{e}^{-\mathrm{i}\beta})}{\zeta - \zeta_c} \right]$$

$$= u_\infty \left(\mathrm{e}^{\mathrm{i}\alpha} + \frac{a \mathrm{e}^{\mathrm{i}\beta}}{\zeta - \zeta_c} \right) \left(1 - \frac{a \mathrm{e}^{-\mathrm{i}(\alpha+\beta)}}{\zeta - \zeta_c} \right)$$

The point $\zeta = -1$ lies on the circle $K(C, a)$, i.e. it holds $a = \zeta_c - (-1) = \zeta_c + 1$ (see Fig. 4.27). By choosing $\beta = \alpha$ we obtain

$$\mathcal{U}(\zeta) = u_\infty \mathrm{e}^{\mathrm{i}\alpha} \left(1 + \frac{\zeta_c + 1}{\zeta - \zeta_c} \right) \left(1 - \frac{\mathrm{e}^{-2\mathrm{i}\alpha}(\zeta_c + 1)}{\zeta - \zeta_c} \right)$$

$$= u_\infty \mathrm{e}^{\mathrm{i}\alpha} \frac{\zeta + 1}{\zeta - \zeta_c} \left(1 - \frac{\mathrm{e}^{-2\mathrm{i}\alpha}(\zeta_c + 1)}{\zeta - \zeta_c} \right)$$

Using (4.60), we find that

$$\hat{\mathcal{U}}(z) = u_\infty \mathrm{e}^{\mathrm{i}\alpha} \frac{\zeta + 1}{\zeta - \zeta_c} \left(1 - \frac{\mathrm{e}^{-2\mathrm{i}\alpha}(\zeta_c + 1)}{\zeta - \zeta_c} \right) \frac{\zeta^2}{(\zeta - 1)(\zeta + 1)}$$

$$= u_\infty \mathrm{e}^{\mathrm{i}\alpha} \frac{1}{\zeta - \zeta_c} \left(1 - \frac{\mathrm{e}^{-2\mathrm{i}\alpha}(\zeta_c + 1)}{\zeta - \zeta_c} \right) \frac{\zeta^2}{\zeta - 1}$$

Thus,

$$\hat{\mathcal{U}}_h = \lim_{z \to -2} \hat{\mathcal{U}}(z) = \frac{u_\infty \mathrm{e}^{\mathrm{i}\alpha}}{2(1 + \zeta_c)} (1 + \mathrm{e}^{-2\mathrm{i}\alpha}) < \infty \tag{4.63}$$

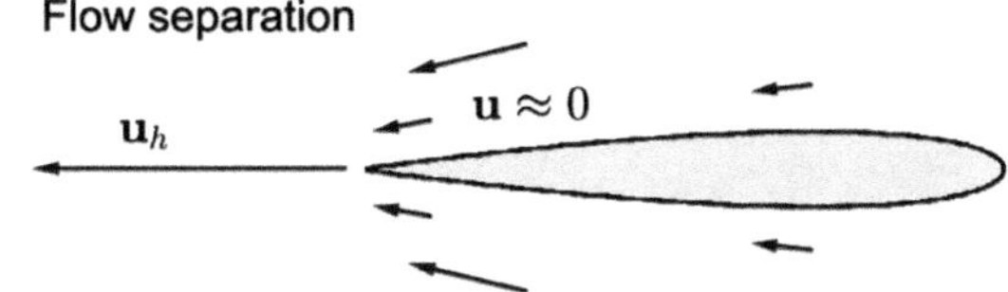

Fig. 4.29 Taking into account the deceleration on the profile, the flow separation indicates a high velocity gradient at the tail.

Considering $e^{i\alpha}(1 + e^{-2i\alpha}) = e^{i\alpha} + e^{-i\alpha} = 2\cos\alpha$, it follows from (4.63) that

$$\hat{u}_h = \frac{u_\infty \cos\alpha}{1 + \zeta_c},$$

i.e. the velocity at the wing tip has no imaginary part. The flow leaves the profile in a horizontal direction, i.e. along the profile axis (Fig. 4.29). $\qquad\square$

We would like to estimate the order of magnitude of the flow velocity when separating from the wing tip. For this purpose, we choose the angle of attack $\alpha < 15°$ and read the center of the circle at $\zeta_c < 0.4$ from Fig. 4.20. By (4.62), we find that $u_h > 0.69\, u_\infty$. In fact, the value is slightly lower, as our model allows the unrealistic boundary condition $\mathbf{u} \neq 0$ and neglects the friction on the wing. However, compared to the decelerated flow at the boundary, the velocity at the tip indicates a high velocity gradient at the tail, which, according to (3.41), causes *increased vortex formation at the wing tip*. It plays a crucial role in the generation of lift.

Using the Kutta-Joukowski theorem (Theorem 4.3), we calculate the force on the wing profile when the flow approaches at an angle α

$$\mathbf{F} = -4a\pi\varrho u_\infty^2 \sin\alpha\, \mathbf{n}, \tag{4.64}$$

Here $\mathbf{n}$ denotes the normal vector to the flow direction.

The formula shows that the profile from Fig. 4.20 cannot generate lift when the flow approaches horizontally. The angle of attack α must be positive, i.e., the wing should be tilted. However, determining α to ensure lift is beyond the scope of our model. This is caused by small vortices that form at an angle of $10° - 15°$ above the nose and cause a loss of lift when detaching. This issue is of great importance for flight safety. We will return to it in the Sects. 5.4 and 6.1.

From the direction of force in (4.64), it becomes apparent that the wing does not develop any drag in the Kutta-Joukowski flow. The result, a consequence of the inviscid-flow approximation, is in astonishing agreement with experiments. Wind tunnel measurements showed that a wing profile of type NACA 64(4)-421 (see Fig. 4.30) has about the same drag as a wire whose diameter is only 0.6% of the profile width ([1], p. 5).

Let us summarize the most important relations that result from (4.53), (4.48) as well as Problem 4.27 and the Joukowski condition.

Fig. 4.30 NACA 64(4)-421.

> **Model equations for calculating lift on the wing profile**
> *Circular potential* for $K(C, a)$ with centre coordinates ζ_c
>
> $$w(\zeta) = u_\infty \left[(\zeta - \zeta_c) e^{i\alpha} + \frac{a^2 e^{-i\alpha}}{\zeta - \zeta_c} \right] - \frac{\Gamma i}{2\pi} \ln(\zeta - \zeta_c), \quad |\zeta - \zeta_c| \geq a \qquad (4.65)$$
>
> *Coordinate transformation* into the wing profile (Joukowski transformation)
>
> $$z = \zeta + \frac{1}{\zeta} \qquad (4.66)$$
>
> *Circulation* to guarantee the Joukowski condition for a given angle of attack α
>
> $$\Gamma = -4a\pi u_\infty \sin \alpha \qquad (4.67)$$

4.8   The Paradox of d'Alembert

The Kutta-Joukowski formula (4.13) allows for the lift in a inviscid and vortex-free flow, as long as the circulation (4.10) does not become zero. In this section, we examine how far this statement can be extended to three-dimensional space. We will see that under inviscid and vortex-free conditions, circulation and thus lift disappear. Flight would therefore be impossible.

In the inviscid lift model, vorticity is therefore indispensable. At the same time, we open up a new perspective on circulation. It proves to be a trick to shift vortex effects from the flow area to the boundary.

Theorem 4.9 (Paradox of d'Alembert) *Let $D \subset \mathbb{R}^3$ be a smooth, bounded domain and $\mathbf{u} = \mathbf{u}(\mathbf{x})$ a stationary potential flow with constant density in the exterior domain $\mathbb{R}^3 \setminus D$. Furthermore, we assume a constant flow at a sufficient distance from D, i.e. it holds that $\lim_{\|\mathbf{x}\| \to \infty} \mathbf{u}(\mathbf{x}) = \mathbf{u}_\infty$.*

Then the flow does not exert any force on the obstacle, i.e. $\mathbf{F} = \int_{\partial D} p\mathbf{n}\, dA = 0$.

Proof For simplification, we set the constant density $\varrho = 1$ and obtain from (3.30), (3.31), (3.32) the stationary Euler equations

$$(\mathbf{u} \cdot \nabla)\mathbf{u} + \nabla p = 0 \quad \text{in } \mathbb{R}^3 \setminus D \qquad (4.68)$$

$$\operatorname{div} \mathbf{u} = 0 \quad \text{in } \mathbb{R}^3 \setminus D \tag{4.69}$$

$$\operatorname{rot} \mathbf{u} = 0 \quad \text{in } \mathbb{R}^3 \setminus D \tag{4.70}$$

and the boundary condition $\mathbf{u} \cdot \mathbf{n} = 0$ on ∂D. Here, $\mathbf{n}$ denotes the outer unit normal vector on ∂D.

In the first step, we look for a potential. Our exterior region is simply connected. Owing to $\operatorname{rot}(\mathbf{u} - \mathbf{u}_\infty) = 0$ and Lemma 1.1, there exists a potential ϕ with

$$\mathbf{u} - \mathbf{u}_\infty = \nabla \phi \qquad \text{in } \mathbb{R}^3 \setminus D \tag{4.71}$$

From $\lim_{\|\mathbf{x}\| \to \infty} \nabla \phi = \lim_{\|\mathbf{x}\| \to \infty} (\mathbf{u} - \mathbf{u}_\infty) = 0$ we find that

$$\lim_{\|\mathbf{x}\| \to \infty} \phi = \phi_\infty$$

By applying the divergence operator, it follows from (4.71)

$$\Delta \phi = 0 \qquad \text{in } \mathbb{R}^3 \setminus D \tag{4.72}$$

and as a boundary condition we obtain $(\mathbf{u}_\infty + \nabla \phi) \cdot \mathbf{n} = 0$ or

$$\frac{d\phi}{d\mathbf{n}} = -\mathbf{u}_\infty \cdot \mathbf{n} = \text{const} \qquad \text{on } \partial D \tag{4.73}$$

By $\hat{\phi}$ we denote the solution of the exterior Neumann problem

$$\Delta \hat{\phi} = 0 \qquad \text{in } \mathbb{R}^3 \setminus D$$

$$\frac{d\hat{\phi}}{d\mathbf{n}} = -\mathbf{u}_\infty \cdot \mathbf{n} \qquad \text{on } \partial D$$

$$\lim_{\|\mathbf{x}\| \to \infty} \hat{\phi}(\mathbf{x}) = 0$$

The existence and uniqueness of the solution is shown in [3]. It can be represented as the sum of the potentials of the simple layer and the double layer (see [3]).

$$\hat{\phi}(\mathbf{x}) = \int_{\partial D} \hat{\phi}(\mathbf{y}) \frac{dG(\mathbf{x}, \mathbf{y})}{d\mathbf{n}_y} \, dA_y - \int_{\partial D} G(\mathbf{x}, \mathbf{y}) \frac{d\hat{\phi}}{d\mathbf{n}_y}(\mathbf{y}) \, dA_y$$

Here, $G(\mathbf{x}, \mathbf{y}) = -\frac{1}{4\pi \|\mathbf{x} - \mathbf{y}\|}$ denotes the Green's function. In the following, we use the notations

$$\mathbf{r} = \mathbf{y} - \mathbf{x}, \quad r = \|\mathbf{y} - \mathbf{x}\|, \quad \mathbf{e}_r = \frac{\mathbf{r}}{r}$$

The potential ϕ from (4.71), (4.73) differs from the solution of the exterior Neumann problem $\hat{\phi}$ by an additive constant ϕ_∞, i.e. it holds

$$\phi(\mathbf{x}) = \phi_\infty + \int_{\partial D} \phi(\mathbf{y}) \frac{dG(\mathbf{x}, \mathbf{y})}{d\mathbf{n}_y} \, dA_y - \int_{\partial D} G(\mathbf{x}, \mathbf{y}) \frac{d\phi}{d\mathbf{n}_y}(\mathbf{y}) \, dA_y$$

If we insert the boundary conditions, we obtain

$$\phi(\mathbf{x}) = \phi_\infty + \int_{\partial D} \phi(\mathbf{y}) \frac{dG(\mathbf{x}, \mathbf{y})}{d\mathbf{n}_y} \, dA_y + \int_{\partial D} G(\mathbf{x}, \mathbf{y}) \, (\mathbf{u}_\infty \cdot \mathbf{n}_y) \, dA_y$$

$$= \phi_\infty + \frac{1}{4\pi} \int_{\partial D} \phi(\mathbf{y})(\mathbf{n}_y \cdot \mathbf{r}) \, r^{-3} \, dA_y - \frac{1}{4\pi} \int_{\partial D} r^{-1}(\mathbf{u}_\infty \cdot \mathbf{n}_y) \, dA_y,$$

where in the last step we replaced the normal derivative of the Green's function (1.31).

Now we consider the obstacle D as a disturbance of the inflow. Using the symbol $O(\cdot)$ (see Sect. A.1), our goal is to prove the formulas

$$\mathbf{u}(\mathbf{x}) = \mathbf{u}_\infty + O(\|\mathbf{x}\|^{-3}), \quad p(\mathbf{x}) = p_\infty + O(\|\mathbf{x}\|^{-3}) \qquad \text{for } \|\mathbf{x}\| \to \infty,$$

from which the inequalities

$$\|\mathbf{u}(\mathbf{x}) - \mathbf{u}_\infty\| \le \frac{\text{const}}{\|\mathbf{x}\|^3}, \quad |p(\mathbf{x}) - p_\infty| \le \frac{\text{const}}{\|\mathbf{x}\|^3}$$

result. For the gradient of the potential, we get

$$\nabla\phi(\mathbf{x}) = \frac{1}{4\pi} \nabla \int_{\partial D} \phi(\mathbf{y})(\mathbf{n}_y \cdot \mathbf{r}) \, r^{-3} \, dA_y - \frac{1}{4\pi} \nabla \int_{\partial D} r^{-1}(\mathbf{u}_\infty \cdot \mathbf{n}_y) \, dA_y$$

$$\nabla\phi(\mathbf{x}) = \frac{1}{4\pi} \int_{\partial D} \phi(\mathbf{y}) \, \nabla_x(\mathbf{n}_y \cdot \mathbf{r}) \, r^{-3} \, dA_y$$

$$+ \frac{1}{4\pi} \int_{\partial D} \phi(\mathbf{y})(\mathbf{n}_y \cdot \mathbf{r}) \, \nabla_x r^{-3} \, dA_y \qquad (4.74)$$

$$- \frac{1}{4\pi} \int_{\partial D} \nabla_x r^{-1}(\mathbf{u}_\infty \cdot \mathbf{n}_y) \, dA_y$$

Using the coordinates $\mathbf{n}_y = (n_i)$ and $\mathbf{r} = (x_i - y_i)$, we find that

$$\partial_j(n_i(x_i - y_i)) = n_i \, \partial_j x_i = n_i \delta_{ij} = n_j$$

and therefore

$$\nabla_x(\mathbf{n}_y \cdot \mathbf{r}) = \mathbf{n}_y \qquad (4.75)$$

Furthermore, we have

$$\mathbf{n}_y \cdot \mathbf{r} = r\,(\mathbf{n}_y \cdot \mathbf{e}_r) \tag{4.76}$$

and by (1.30)

$$\nabla_x r^{-1} = -r^{-2}\,\mathbf{e}_r, \qquad \nabla_x r^{-3} = -3r^{-4}\,\mathbf{e}_r \tag{4.77}$$

If we substitute (4.75), (4.76), (4.77) into (4.74), we obtain

$$\begin{aligned}
\nabla\phi(\mathbf{x}) = {} & \frac{1}{4\pi}\int_{\partial D}\phi(\mathbf{y})\,\mathbf{n}_y\,r^{-3}\,\mathrm{d}A_y \\
& -\frac{3}{4\pi}\int_{\partial D}\phi(\mathbf{y})\,(\mathbf{n}_y\cdot\mathbf{e}_r)\,r^{-3}\,\mathbf{e}_r\,\mathrm{d}A_y \\
& +\frac{1}{4\pi}\int_{\partial D}\frac{\mathbf{e}_r}{r^2}\,(\mathbf{u}_\infty\cdot\mathbf{n}_y)\,\mathrm{d}A_y
\end{aligned} \tag{4.78}$$

The first two integrals are of order $O(\|\mathbf{x}\|^{-3})$. To evaluate the last integral, we set

$$\mathbf{u}_\infty = \begin{pmatrix} u_1 \\ u_2 \\ u_3 \end{pmatrix},\quad
\mathbf{x} = \begin{pmatrix} x_1 \\ x_2 \\ x_3 \end{pmatrix},\quad
\mathbf{y} = \begin{pmatrix} y_1 \\ y_2 \\ y_3 \end{pmatrix},\quad
\mathbf{a} = \frac{\mathbf{e}_r}{r^2} = \frac{1}{\|\mathbf{x}-\mathbf{y}\|^3}\begin{pmatrix} x_1 - y_1 \\ x_2 - y_2 \\ x_3 - y_3 \end{pmatrix}$$

and use Corollary 1.1

$$\int_{\partial D}\mathbf{a}(\mathbf{n}\cdot\mathbf{u}_\infty)\,\mathrm{d}S_y = -\int_D [\mathbf{a}(\operatorname{div}\mathbf{u}_\infty) + (\mathbf{u}_\infty\cdot\nabla)\mathbf{a}]\,\mathrm{d}V_y$$

Since $\mathbf{u}_\infty$ is constant, it follows that $\operatorname{div}\mathbf{u}_\infty = 0$ and

$$\int_{\partial D}\mathbf{a}(\mathbf{n}\cdot\mathbf{u}_\infty)\,\mathrm{d}S_y = -\int_D (\mathbf{u}_\infty\cdot\nabla)\mathbf{a}\,\mathrm{d}V_y \tag{4.79}$$

We use (2.3) to write the directional derivative in the form

$$(\mathbf{u}_\infty\cdot\nabla)\mathbf{a} = \begin{pmatrix}
\sum_{i=1}^{3} u_i\,\partial_{y_i}\frac{x_1-y_1}{\|\mathbf{x}-\mathbf{y}\|^3} \\
\sum_{i=1}^{3} u_i\,\partial_{y_i}\frac{x_2-y_2}{\|\mathbf{x}-\mathbf{y}\|^3} \\
\sum_{i=1}^{3} u_i\,\partial_{y_i}\frac{x_3-y_3}{\|\mathbf{x}-\mathbf{y}\|^3}
\end{pmatrix}$$

For short, we denote the scalar product by

$$(\mathbf{x}-\mathbf{y})\cdot(\mathbf{x}-\mathbf{y}) = \langle \mathbf{x}-\mathbf{y},\,\mathbf{x}-\mathbf{y}\rangle$$

To estimate the orders, we get

$$\partial_{y_i} \frac{x_i - y_i}{\|\mathbf{x} - \mathbf{y}\|^3} = \partial_{y_i}[(x_i - y_i)\langle \mathbf{x} - \mathbf{y}, \mathbf{x} - \mathbf{y}\rangle^{-3/2}]$$

$$= \frac{3(x_i - y_i)^2}{\|\mathbf{x} - \mathbf{y}\|^5} + \frac{1}{\|\mathbf{x} - \mathbf{y}\|^3} = O(\|\mathbf{x}\|^{-3})$$

Similarly, for $i \neq j$, it follows that

$$\partial_{y_i} \frac{x_j - y_j}{\|\mathbf{x} - \mathbf{y}\|^3} = (x_j - y_j)\partial_{y_i}[\langle \mathbf{x} - \mathbf{y}, \mathbf{x} - \mathbf{y}\rangle^{-3/2}]$$

$$= \frac{3(x_i - y_i)(x_j - y_j)}{\|\mathbf{x} - \mathbf{y}\|^5} = O(\|\mathbf{x}\|^{-3})$$

Hence each component of the vector $(\mathbf{u}_\infty \cdot \nabla)\mathbf{a}$ is of order $O(\|\mathbf{x}\|^{-3})$ and we obtain from (4.79)

$$\int_{\partial D} \mathbf{a}(\mathbf{u}_\infty \cdot \mathbf{n})\, \mathrm{d}A_y = O(\|\mathbf{x}\|^{-3})$$

The same applies to the other integrals in (4.78). Thus, $\nabla\phi(\mathbf{x}) = O(\|\mathbf{x}\|^{-3})$, and consequently

$$\mathbf{u}(\mathbf{x}) = \mathbf{u}_\infty + \nabla\phi(\mathbf{x}) = \mathbf{u}_\infty + O(\|\mathbf{x}\|^{-3}) \tag{4.80}$$

To examine the pressure, we evaluate

$$\|\mathbf{u}(\mathbf{x})\|^2 = (\mathbf{u}_\infty + O(\|\mathbf{x}\|^{-3}))^2 = \|\mathbf{u}_\infty\|^2 + 2\|\mathbf{u}_\infty\|\, O(\|\mathbf{x}\|^{-3}) + O(\|\mathbf{x}\|^{-3})^2$$

$$= \|\mathbf{u}_\infty\|^2 + O(\|\mathbf{x}\|^{-3})$$

Applying Bernoulli's theorem (Theorem 3.2, vortex-free case) in $\mathbb{R}^3 \setminus D$

$$\frac{1}{2}\|\mathbf{u}(\mathbf{x})\|^2 + \frac{p(\mathbf{x})}{\varrho} = \frac{1}{2}\|\mathbf{u}_\infty\|^2 + \frac{p_\infty}{\varrho}$$

$$\frac{\varrho}{2}\left[\|\mathbf{u}_\infty\|^2 + O(\|\mathbf{x}\|^{-3})\right] + p(\mathbf{x}) = \frac{\varrho}{2}\|\mathbf{u}_\infty\|^2 + p_\infty,$$

we find for the pressure

$$p(\mathbf{x}) = p_\infty + O(\|\mathbf{x}\|^{-3}) \tag{4.81}$$

Now let Σ be a closed spherical surface with centre at zero and radius R, which completely contains D (Fig. 4.31).

Fig. 4.31 The spherical
surface Σ encloses the
obstacle D.

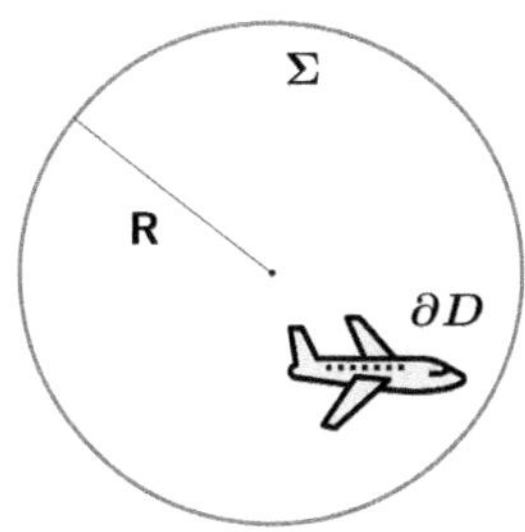

By W we denote the region enclosed by ∂D and Σ. Using the integral theorems, we show the equation

$$\int_{\partial W} [p\,\mathbf{n} + \varrho(\mathbf{u} \cdot \mathbf{n})\,\mathbf{u}]\,\mathrm{d}S = 0 \tag{4.82}$$

Using (1.25), we find that

$$\int_{\partial W} p\,\mathbf{n}\,\mathrm{d}S = \int_W \operatorname{grad} p\,\mathrm{d}V$$

Further, applying (1.24), we obtain

$$\int_{\partial W} \varrho(\mathbf{u} \cdot \mathbf{n})\mathbf{u}\,\mathrm{d}S = \int_W [(\varrho\mathbf{u})\operatorname{div}\mathbf{u} + \varrho(\mathbf{u} \cdot \nabla)\,\mathbf{u}]\,\mathrm{d}V$$

$$= \int_W \varrho(\mathbf{u} \cdot \nabla)\,\mathbf{u}\,\mathrm{d}V \qquad \text{using div }\mathbf{u} = 0$$

In summary, taking into account (4.68), we have

$$\int_{\partial W} [p\,\mathbf{n} + \varrho(\mathbf{u} \cdot \mathbf{n})\mathbf{u}]\,\mathrm{d}S = \int_W [\operatorname{grad} p + \varrho(\mathbf{u} \cdot \nabla)\,\mathbf{u}]\,\mathrm{d}V = 0$$

Since $\partial W = \Sigma \cup \partial D$, the integral (4.82) can be decomposed and we obtain

$$\int_{\partial D} (p\,\mathbf{n} + \varrho(\mathbf{u} \cdot \mathbf{n})\,\mathbf{u})\,\mathrm{d}S = -\int_{\Sigma} (p\,\mathbf{n} + \varrho(\mathbf{u} \cdot \mathbf{n})\,\mathbf{u})\,\mathrm{d}S \tag{4.83}$$

The force acting on D is

$$\mathbf{F} = \int_{\partial D} p\,\mathbf{n}\,\mathrm{d}S \tag{4.84}$$

$$= \int_{\partial D} (p\,\mathbf{n} + \varrho(\mathbf{u} \cdot \mathbf{n})\,\mathbf{u})\,\mathrm{d}S \quad \text{because } \mathbf{u} \cdot \mathbf{n} = 0 \text{ on } \partial D$$

Using (4.83), we can express $\mathbf{F}$ as an integral over Σ.

$$\mathbf{F} = -\int_\Sigma (p\,\mathbf{n} + \varrho(\mathbf{u}\cdot\mathbf{n})\,\mathbf{u})\,\mathrm{d}S$$

By (4.80) and (4.81), we find that

$$p\,\mathbf{n} = [p_\infty + O(\|\mathbf{x}\|^{-3})]\,\mathbf{n} = p_\infty\,\mathbf{n} + O(\|\mathbf{x}\|^{-3})$$

$$\mathbf{u}\cdot\mathbf{n} = [\mathbf{u}_\infty + O(\|\mathbf{x}\|^{-3})]\cdot\mathbf{n} = \mathbf{u}_\infty\cdot\mathbf{n} + O(\|\mathbf{x}\|^{-3})$$

$$\varrho(\mathbf{u}\cdot\mathbf{n})\,\mathbf{u} = \varrho[\mathbf{u}_\infty\cdot\mathbf{n} + O(\|\mathbf{x}\|^{-3})]\,[\mathbf{u}_\infty + O(\|\mathbf{x}\|^{-3})]$$

$$= \varrho(\mathbf{u}_\infty\cdot\mathbf{n})\,\mathbf{u}_\infty + O(\|\mathbf{x}\|^{-3})$$

and obtain

$$\mathbf{F} = -\int_\Sigma [p_\infty\,\mathbf{n} + O(\|\mathbf{x}\|^{-3}) + \varrho(\mathbf{u}_\infty\cdot\mathbf{n})\,\mathbf{u}_\infty + O(\|\mathbf{x}\|^{-3})]\,\mathrm{d}S$$

$$= -\int_\Sigma [p_\infty\,\mathbf{n} + \varrho(\mathbf{u}_\infty\cdot\mathbf{n})\,\mathbf{u}_\infty]\,\mathrm{d}S - \int_\Sigma O(\|\mathbf{x}\|^{-3})\,\mathrm{d}S \qquad (4.85)$$

We show that the first integral of (4.85) vanishes.

$$\int_\Sigma [p_\infty\,\mathbf{n} + \varrho(\mathbf{u}_\infty\cdot\mathbf{n})\,\mathbf{u}_\infty]\,\mathrm{d}S = p_\infty\int_\Sigma \mathbf{n}\,\mathrm{d}S + \varrho\,\mathbf{u}_\infty\int_\Sigma (\mathbf{u}_\infty\cdot\mathbf{n})\,\mathrm{d}S$$

$$= p_\infty\int_\Sigma \mathbf{n}\,\mathrm{d}S + \varrho\,\mathbf{u}_\infty\left(\mathbf{u}_\infty\cdot\int_\Sigma \mathbf{n}\,\mathrm{d}S\right)$$

$$= 0 \qquad \text{by Problem 1.7}$$

By (4.85) this yields

$$\mathbf{F} = -\int_\Sigma O(\|\mathbf{x}\|^{-3})\,\mathrm{d}S$$

$$= O(R^{-3})\,4\pi R^2 \to 0 \quad \text{for } R \to \infty$$

$$\square$$

In 1752, d'Alembert discovered the special case of no drag using energetic considerations. The result is in obvious contradiction to reality and indicates a problem in the model. Consequently, vortex formation plays a key role in the generation of lift.

Why does the more realistic three-dimensional model fail, while the less realistic two-dimensional model provides useful formulas for lift? The paradox is based on the topological difference of domains in $\mathbb{R}^2$ and $\mathbb{R}^3$. As Example 1.1 shows, domains around aircraft and birds are of type $\mathbb{R}^3 \setminus D$, i.e. simply connected. Here, the assumed vortex-free state $\operatorname{rot} \mathbf{u} = 0$ implies, according to Lemma 1.1, the existence of a potential ϕ, i.e. we have $\mathbf{u} = \nabla \phi$. Therefore, the integral along a curve from P_1 to P_2 is path independent

$$\int_{P_1}^{P_2} \mathbf{u}\, d\mathbf{x} = \phi(P_2) - \phi(P_1)$$

In particular, we obtain $\Gamma = \oint_C \mathbf{u}\, d\mathbf{x} = 0$ around every closed curve C, i.e. circulation and lift vanish simultaneously in the three-dimensional case. In contrast, the domain $\mathbb{R}^2 \setminus D$ is multiply connected. Despite the lack of vorticity, there is no potential, i.e. the circulation does not necessarily have to vanish.

References and Further Reading

1. Jones R.T.: *Theory of Wings.* Princeton University Press, Princeton (2014)
2. Maklakov D.V.: *Analog of the Kutta-Joukowskii theorem for the Helmholtz-Kirchhoff flow past a profile*, Doklady Physics 56(11), 573–576 (2011); translation from Dokl. Akad. Nauk 441(2), 187–190 (2011).
3. Vladimirov W.S.: *Equations of Mathematical Physics.* Mir Publishers, Moscow (1984)

Vorticity is essential for flight. But how do they arise? Before the aircraft takes off, the surrounding air is at rest, i.e. in a vortex-free state. We will prove that the vorticity in the Euler equations can neither arise nor disappear. Thus, the frictionless model cannot generate vorticity and is unsuitable for explaining the take-off process.

5.1 Circulation in the Frictionless Model

In a three-dimensional domain $\Omega \subset \mathbb{R}^3$, we consider an ideal fluid with constant density. It satisfies the non-stationary Euler equations.

$$\frac{\partial \mathbf{u}}{\partial t} + (\mathbf{u} \cdot \nabla)\mathbf{u} + \frac{1}{\varrho}\nabla p = 0 \quad \text{in } \Omega \qquad \text{(equation of motion)} \qquad (5.1)$$

$$\operatorname{div} \mathbf{u} = 0 \quad \text{in } \Omega \qquad \text{(incompressibility)} \qquad (5.2)$$

$$\mathbf{u} \cdot \mathbf{n} = 0 \quad \text{on } \partial\Omega \qquad \text{(boundary condition)} \qquad (5.3)$$

In contrast to the assumptions of the d'Alembert paradox, we insist here neither on stationarity nor on the absence of vorticity.

A curve as part of a fluid is called a *material curve* if it consists of the same particles at all times, i.e. it floats with the flow.

R. Spielmann, *Theoretical Fluid Mechanics*,
https://doi.org/10.1007/978-3-662-72240-4_5

Theorem 5.1 (Kelvin) *Let $C(t)$ be a simple closed, material curve in an ideal fluid. Then it holds*

$$\frac{\mathrm{d}}{\mathrm{d}t}\Gamma_{C(t)} = 0, \tag{5.4}$$

i.e. the circulation $\Gamma_{C(t)}$ is invariant under the flow.

Proof We parametrise $C(t)$ as $\mathbf{x}(\alpha, t)$, $0 \le \alpha \le 1$. Then we obtain

$$\frac{\mathrm{d}}{\mathrm{d}t}\Gamma_{C(t)} = \frac{\mathrm{d}}{\mathrm{d}t}\oint_{C(t)} \mathbf{u}\,\mathrm{d}\mathbf{x} = \frac{\mathrm{d}}{\mathrm{d}t}\int_0^1 \mathbf{u}\cdot\frac{\partial\mathbf{x}}{\partial\alpha}\,\mathrm{d}\alpha = \int_0^1 \left[\frac{\mathrm{d}\mathbf{u}}{\mathrm{d}t}\cdot\frac{\partial\mathbf{x}}{\partial\alpha} + \mathbf{u}\cdot\frac{\mathrm{d}\mathbf{u}}{\mathrm{d}\alpha}\right]\mathrm{d}\alpha$$

By (2.4), we have $\frac{\mathrm{d}\mathbf{u}}{\mathrm{d}t} = \frac{\partial\mathbf{u}}{\partial t} + (\mathbf{u}\cdot\nabla)\mathbf{u}$. Consequently, (5.1) takes the form $\frac{\mathrm{d}\mathbf{u}}{\mathrm{d}t} = -\frac{1}{\varrho}\nabla p$ and we get

$$\frac{\mathrm{d}}{\mathrm{d}t}\Gamma_{C(t)} = \int_0^1\left[-\frac{1}{\varrho}\nabla p\cdot\frac{\partial\mathbf{x}}{\partial\alpha} + \mathbf{u}\cdot\frac{\partial\mathbf{u}}{\partial\alpha}\right]\mathrm{d}\alpha = -\frac{1}{\varrho}\int_0^1 \nabla p\cdot\frac{\partial\mathbf{x}}{\partial\alpha}\,\mathrm{d}\alpha + \int_0^1 \frac{\mathrm{d}}{\mathrm{d}\alpha}\left(\frac{\|\mathbf{u}\|^2}{2}\right)\mathrm{d}\alpha$$

Since $C(t)$ remains closed for all t, it holds that $\mathbf{x}(0, t) = \mathbf{x}(1, t)$ and thus

$$p(\mathbf{x}(0, t), t) = p(\mathbf{x}(1, t), t), \qquad \mathbf{u}(\mathbf{x}(0, t), t) = \mathbf{u}(\mathbf{x}(1, t), t)$$

Therefore,

$$\frac{\mathrm{d}}{\mathrm{d}t}\Gamma_{C(t)} = -\frac{1}{\varrho}\int_0^1 \frac{\mathrm{d}p}{\mathrm{d}\alpha}\,\mathrm{d}\alpha + \frac{1}{2}\int_0^1 \frac{\mathrm{d}}{\mathrm{d}\alpha}\|\mathbf{u}\|^2\,\mathrm{d}\alpha = 0$$

$\square$

Corollary 5.1 *Consider an ideal fluid and suppose that $\operatorname{rot}\mathbf{u} = 0$ at time $t = 0$. Then it follows that*

$$\operatorname{rot}\mathbf{u} = 0 \qquad \textit{for all } t.$$

Proof We consider a simple closed, material curve $C(t)$, that encloses the surface $\Sigma(t)$. Applying the theorems of Stokes (Theorem 1.2) and Kelvin (Theorem 5.1), we find that

$$\int_{\Sigma(t)} \operatorname{rot}\mathbf{u}\,\mathrm{d}S = \oint_{C(t)} \mathbf{u}\,\mathrm{d}\mathbf{x} = \Gamma_{C(t)} = \text{const} \qquad \text{for all } t$$

Taking into account the initial value for $t = 0$, we get

$$\int_{\Sigma(t)} \operatorname{rot} \mathbf{u} \, dS = 0 \qquad \text{for all } t$$

Since $C(t)$ can be chosen arbitrarily, the assertion follows. $\square$

Now it becomes clear why the take-off of an aircraft cannot be explained in the frictionless model. Since the flow was at rest before take-off and no vortex can form in the ideal fluid, no lift is possible here.

5.2   Generation of Vorticity at a Boundary Layer

Friction is essential for the creation of lift. We show how friction in the boundary layer creates a vortex field, that enables lift. The principle becomes clear when we examine the deceleration at the boundary.

Problem 5.1 We consider a two-dimensional flow along an infinite plate, whose velocity vanishes at the boundary. The particles move parallel to the boundary, with their velocity remaining constant on each streamline (Fig. 5.1). Explain why the flow is vortex-free.

For further analysis we use the Navier-Stokes equations for a homogeneous, incompressible fluid.

$$\frac{\partial \mathbf{u}}{\partial t} + (\mathbf{u} \cdot \nabla)\mathbf{u} + \frac{1}{\varrho}\nabla p = \nu \, \Delta \mathbf{u} \quad \text{in } \mathbb{R}^3 \setminus D \quad \text{(equation of motion)} \quad (5.5)$$

$$\operatorname{div} \mathbf{u} = 0 \qquad \text{in } \mathbb{R}^3 \setminus D \quad \text{(incompressibility)} \qquad (5.6)$$

$$\mathbf{u} = 0 \qquad \text{on } \partial D \qquad \text{(boundary condition)} \qquad (5.7)$$

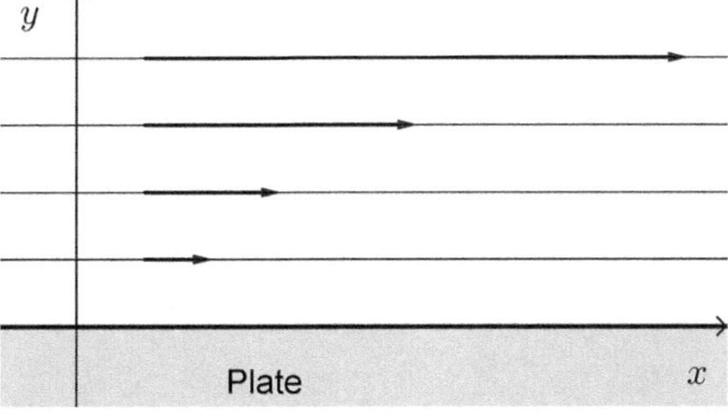

Fig. 5.1 A flow runs parallel to a plate and is completely slowed down at the boundary. The particle velocities on the streamlines are illustrated by the direction and length of the arrows.

Here, ν denotes the viscosity. For simplicity we restrict ourselves to the two-dimensional flow $\mathbf{u} = (u(y, t), 0)$ under the conditions

$$u(0, t) = 0 \qquad \text{no flow on the plate}$$
$$u(\infty, t) = U \qquad \text{undisturbed flow at a sufficient distance parallel to plate}$$
$$\nabla p = 0 \qquad \text{constant pressure}$$

Since $\mathbf{u}$ is independent of x, we have

$$(\mathbf{u} \cdot \nabla)\mathbf{u} = u \cdot \frac{\partial \mathbf{u}}{\partial x} + 0 \cdot \frac{\partial \mathbf{u}}{\partial y} = \mathbf{0}$$

and the Navier-Stokes equation (5.5) reduces to the unsteady equation

$$\frac{\partial u}{\partial t} = \nu \frac{\partial^2 u}{\partial y^2} \tag{5.8}$$

Now we carry out a scaling. Using the transformation

$$y' = y/L, \qquad t' = t/T$$

we obtain

$$\frac{\partial u}{\partial t'} = \nu \frac{T}{L^2} \frac{\partial^2 u}{\partial y'^2},$$

whereby the boundary conditions remain unchanged.

$$u(0, t') = 0, \qquad u(\infty, t') = U$$

Since the equation remains the same for $L^2 = T$ and its solution is unique, it follows that

$$u(y', t') = u\left(\frac{y}{L}, \frac{t}{T}\right) = u(y, t)$$

In particular, for $T = t$ and $L = \sqrt{t}$,

$$u\left(\frac{y}{\sqrt{t}}, 1\right) = u(y, t)$$

The solution can be expressed by a single variable $\eta = y/\sqrt{t}$ and we find that

$$u(y, t) = u\left(\frac{y}{\sqrt{t}}, 1\right) = U f(\eta)$$

as well as

$$f(\infty) = 1, \qquad f(0) = 0.$$

We determine f as the solution of an ordinary differential equation. From (5.8) we obtain

$$\frac{\partial(Uf)}{\partial t} = \nu \frac{\partial^2(Uf)}{\partial y^2}$$

$$\frac{\partial f}{\partial t} = \nu \frac{\partial^2 f}{\partial y^2}$$

$$\frac{\partial f}{\partial \eta}\frac{\partial \eta}{\partial t} = \nu \frac{\partial}{\partial y}\left(\frac{\partial f}{\partial \eta}\frac{\partial \eta}{\partial y}\right) = \nu\left(\frac{\partial^2 f}{\partial \eta^2}\left(\frac{\partial \eta}{\partial y}\right)^2 + \frac{\partial f}{\partial \eta}\frac{\partial^2 \eta}{\partial y^2}\right)$$

Using $\eta = \frac{yt^{-1/2}}{2\sqrt{\nu}}$, we find that

$$\frac{\partial \eta}{\partial y} = \frac{t^{-1/2}}{2\sqrt{\nu}}, \quad \left(\frac{\partial \eta}{\partial y}\right)^2 = \frac{1}{4\nu t}, \quad \frac{\partial \eta}{\partial t} = -\frac{t^{-3/2}y}{4\sqrt{\nu}}, \quad \frac{\partial^2 \eta}{\partial y^2} = 0$$

Thus, we obtain the ordinary differential equation

$$-f'\frac{t^{-3/2}y}{4\sqrt{\nu}} = \nu f'' \frac{1}{4\nu t}$$

$$0 = 2f'\eta + f''$$

Taking into account the boundary conditions, we get

$$f'(\eta) = c\mathrm{e}^{-\eta^2}, \quad c = \text{const} \qquad \text{as a result of the first integration}$$

$$f(\eta) = \frac{2}{\sqrt{\pi}}\int_0^\eta \mathrm{e}^{-s^2}\,\mathrm{d}s \qquad \text{as a result of further integration}$$

Using the *Gaussian error function* $\operatorname{erf}(\eta) = \frac{1}{\sqrt{\pi}}\int_0^\eta \mathrm{e}^{-s^2}\,\mathrm{d}s$, we get $f(\eta) = 2\operatorname{erf}(\eta)$. Thus, our solution is

$$\mathbf{u} = (u(y,t),0) \quad \text{with } u(y,t) = \frac{2U}{\sqrt{\pi}}\int_0^{y/(2\sqrt{\nu t})} \mathrm{e}^{-s^2}\,\mathrm{d}s = 2U\operatorname{erf}\left(\frac{y}{2\sqrt{\nu t}}\right)$$

The flow over the plate generates a vorticity.

$$\operatorname{rot}\mathbf{u} = \frac{\partial v}{\partial x} - \frac{\partial u}{\partial y} = -\frac{\partial u}{\partial y} = -\frac{U}{\sqrt{\pi}}\mathrm{e}^{-y^2/(4\sqrt{\nu t})}\frac{1}{\sqrt{\nu t}} \neq 0$$

5.3   The Boundary Layer Equation

Since the Navier-Stokes equations (3.20), (3.21), (3.22) could only be solved numerically so far, we are looking for a simple approximation. As the main generator of vorticity, the boundary layer is of particular interest. The following model reduces (3.20) to the essential terms in the boundary region. Additionally, it allows to estimate the thinness of the boundary layer. This justifies Joukowski's inviscid-flow approach, which considers vorticity only in the form of boundary circulation.

In analogy with (3.23) we use the variable transformation

$$x' = \frac{x}{L}, \; y' = \frac{y}{\delta}, \; u' = \frac{u}{U}, \; v' = \frac{v}{U}\frac{L}{\delta}, \; p' = \frac{p}{\varrho U^2}, \; t' = \frac{tU}{L}, \tag{5.9}$$

where the constants are summarized in the Reynolds number $\mathrm{Re} = UL/\nu$.

Problem 5.2 Prove that (5.9) transforms the two-dimensional Navier-Stokes equation (3.20) into the following form.

$$\partial_{t'}u' + u'\partial_{x'}u' + v'\partial_{y'}u' = -\partial_{x'}p' + \frac{1}{\mathrm{Re}}\left[\frac{\partial^2 u'}{\partial(x')^2} + \left(\frac{L}{\delta}\right)^2\frac{\partial^2 u'}{\partial(y')^2}\right] \tag{5.10}$$

$$\partial_{t'}v' + u'\partial_{x'}v' + v'\partial_{y'}v' = -\left(\frac{L}{\delta}\right)^2\partial_{y'}p' + \frac{1}{\mathrm{Re}}\left[\frac{\partial^2 v'}{\partial(x')^2} + \left(\frac{L}{\delta}\right)^2\frac{\partial^2 v'}{\partial(y')^2}\right] \tag{5.11}$$

Theorem 5.2 (Boundary Layer Equation by Prandtl) *Consider a two-dimensional flow of low viscosity along a horizontal plate (Fig. 5.2). Near the boundary, except at the tip of the plate, the Navier-Stokes equations (3.20), (3.21), (3.22) simplify to*

$$\partial_t u + u\partial_x u + v\partial_y u = -\frac{1}{\varrho}\partial_x p + \nu\frac{\partial^2 u}{\partial y^2} \tag{5.12}$$

$$\partial_y p = 0 \tag{5.13}$$

$$\partial_x u + \partial_y v = 0 \tag{5.14}$$

$$u = v = 0 \quad \textit{for } y = 0 \tag{5.15}$$

To adapt (5.12)–(5.15) to curved boundaries, we use the following notations.

$$\begin{array}{ll} x & \text{path length along the boundary} \\ y & \text{distance along the normal} \\ \kappa & \text{curvature} \end{array}$$

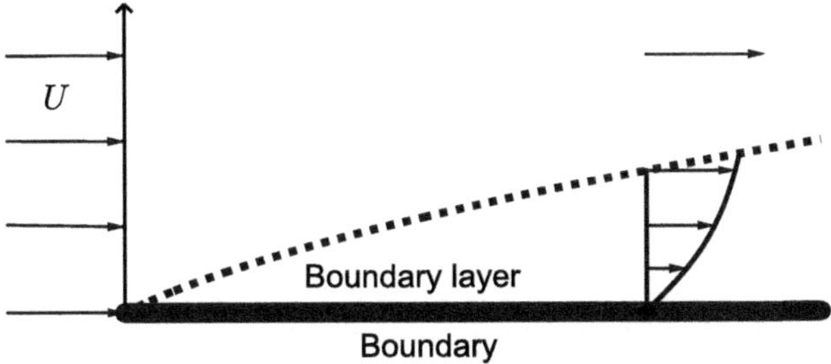

Fig. 5.2 Boundary layer on a plate with horizontal inflow.

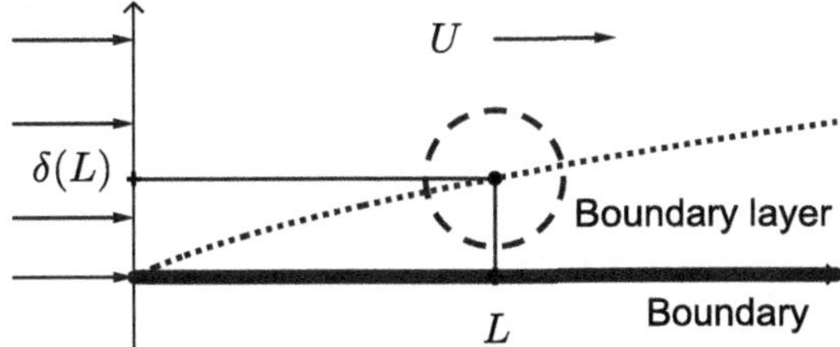

Fig. 5.3 In the boundary layer, the Navier-Stokes equation (3.20) can be transformed into the form (5.12), (5.13). For this purpose, we apply the transformation (5.9) in a neighbourhood of $(L, \delta(L))$.

As long as the curvature is not too strong and does not change drastically, i.e. for small $\kappa\delta$, $\delta L\, \mathrm{d}\kappa/\mathrm{d}x$, the equations remain valid (see [3, 7]).

Proof In the two-dimensional case, the Navier-Stokes equation (3.20) is written as

$$\partial_t u + u\partial_x u + v\partial_y u = -\frac{1}{\varrho}\partial_x p + \nu\Delta u$$

$$\partial_t v + u\partial_x v + v\partial_y v = -\frac{1}{\varrho}\partial_y p + \nu\Delta v$$

From Problem 5.2 we obtain the Eqs. (5.10) and (5.11), where the constants of the underlying transformation (5.9) are chosen as follows (Fig. 5.3):

L	horizontal length scale
U	Flow speed at a sufficient distance from the plate
$\delta \ll L$	Thickness of the boundary layer at $x = L$

It remains to determine which terms of (5.10), (5.11) provide the essential contribution to the flow behaviour of the boundary layer.

In the Navier-Stokes equations, the force $(\mathbf{u} \cdot \nabla)\mathbf{u}$ is referred to as the *inertia term*, while $\nu\Delta\mathbf{u}$ is called the *friction term*. Their influence on the flow behaviour depends on the location considered. While the friction term is of crucial importance directly at the boundary, the inertia term dominates the flow of weakly viscous fluids far from the boundary (see Sect. 3.6).

At a sufficient distance from the tip we consider a point $(L, \delta(L))$ (see Fig. 5.3).

In a suitable neighbourhood of $(L, \delta(L))$ we obtain

$$\text{for the inertia term} \qquad u\frac{\partial u}{\partial x} \sim U\frac{U}{L}$$

as well as[1]

$$\text{for the friction term} \qquad \nu\frac{\partial^2 u}{\partial y^2} \sim \nu\frac{U/\delta}{\delta} = \nu\frac{U}{\delta^2}$$

In a strip around the dotted curve of Fig. 5.3, which exits at the tip, the friction forces are balanced with the inertia forces and the pressure gradient. Thus,

$$\frac{\text{friction term}}{\text{inertia term}} \sim \frac{\nu U/\delta^2}{U^2/L} = O(1)$$

Therefore, we have

$$\frac{1}{\text{Re}}\left(\frac{L}{\delta}\right)^2 = O(1), \qquad \delta = O(\text{Re}^{-1/2}) \tag{5.16}$$

Multiplying of (5.11) by $\frac{1}{\text{Re}}$ yields

$$\frac{1}{\text{Re}}(\partial_{t'}v' + u'\partial_{x'}v' + v'\partial_{y'}v') = -\frac{1}{\text{Re}}\left(\frac{L}{\delta}\right)^2\partial_{y'}p' + \frac{1}{\text{Re}^2}\frac{\partial^2 v'}{\partial(x')^2} + \frac{1}{\text{Re}^2}\left(\frac{L}{\delta}\right)^2\frac{\partial^2 v'}{\partial(y')^2} \tag{5.17}$$

By assumption, ν is small. Thus, we can use the approximation of large Reynolds numbers. Considering (5.16) and $\text{Re} \to \infty$, it follows from (5.10), (5.17) that

$$\partial_{t'}u' + u'\partial_{x'}u' + v'\partial_{y'}u' = -\partial_{x'}p' + \frac{1}{\text{Re}}\left(\frac{L}{\delta}\right)^2\frac{\partial^2 u'}{\partial(y')^2}$$

$$0 = -\frac{1}{\text{Re}}\left(\frac{L}{\delta}\right)^2\partial_{y'}p'$$

By back-transformation, we obtain (5.12) and (5.13). $\square$

---

[1] Since $\frac{\partial^2 u}{\partial x^2} \sim \frac{U/L}{L} \ll \frac{U}{\delta^2} \sim \frac{\partial^2 u}{\partial y^2}$, we can neglect $\frac{\partial^2 u}{\partial x^2}$ in the friction term.

According to (5.13), particles of the boundary layer experience neither acceleration nor deceleration *in the y-direction*. In addition to the laminar case discussed, turbulent boundary layer flows can also occur. When an aircraft takes off, the increasing flow velocity in the boundary layer of the wing causes turbulence to form, which allows flow separation at the wing tip. This process requires high Reynolds numbers and is essential for generating aerodynamic lift.

5.4 Flow Separation

The *Coanda effect* (Fig. 5.4, left) refers to the tendency of a flow to follow along a convex surface. If it separates instead, it is called *flow separation* (Fig. 5.4, right).

In Sect. A.6, we will describe the situation using a complex potential.

Regarding the flow around an airfoil, we speak of *early flow separation* or simply flow separation, when the vortex-rich boundary layer separates before reaching the wing tip. Its outflow reduces the lift.

An early flow separation was shown in Figs. 4.2 and 4.4 at the beginning of Chap. 4 and is the subject of extensive experimental research. We explain the principle of its formation using a simplified approach. Particles flowing against increasing pressure are slowed down, while they are accelerated when the pressure decreases (see (4.6) or Exercise 4.2).

Let us consider a stationary flow around a cylinder, whose boundary flow separates in the rear area (Fig. 5.5).

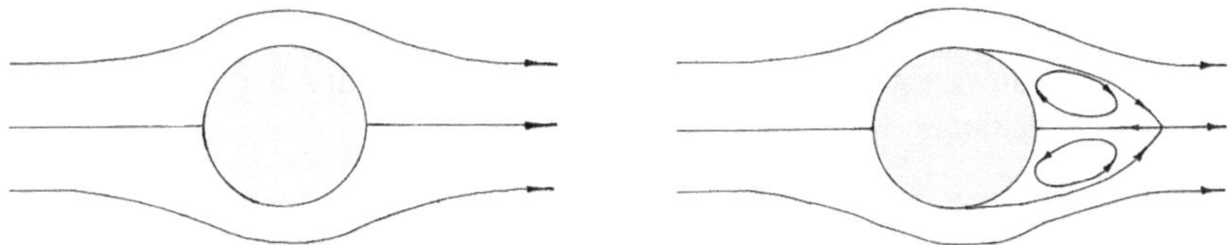

Fig. 5.4 (left) Re $\approx$ 0.01. Coanda effect on a flow around a circular cylinder. (right) Re $\approx$ 20. At higher Reynolds numbers, flow separation occurs.

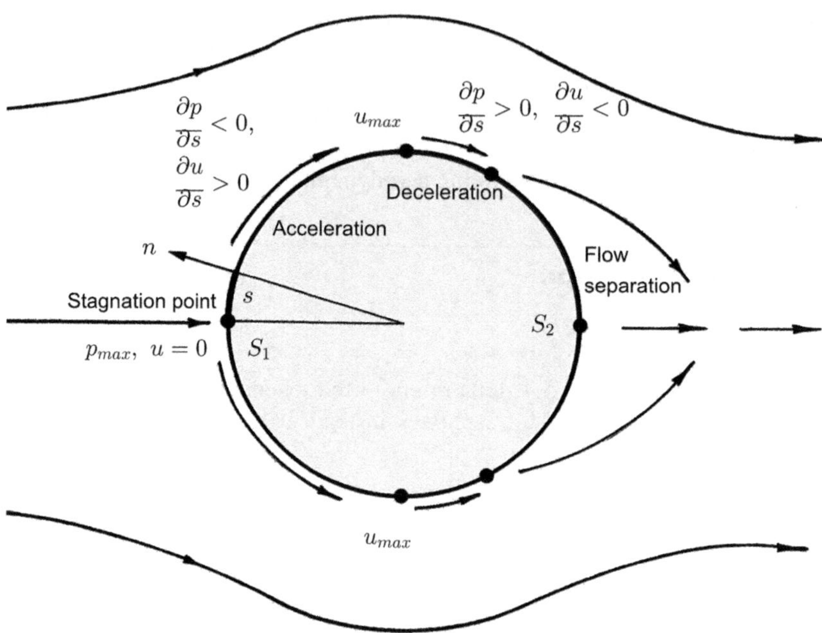

Fig. 5.5 Boundary layer separation in a stationary flow around a cylinder. The stagnation points are located at S_1 and S_2.

Above the cylinder, we describe the flow with an *s-coordinate as the path length* on the cylinder and an *n-coordinate in the normal direction.*

According to the remark after Theorem 5.2, the boundary layer equation retains its validity in curvilinear coordinates. In the *inner part of the boundary layer*, i.e. directly at the boundary, it can be further simplified. Using stationarity and $u \approx 0$, $v \approx 0$, we obtain from (5.12)

$$0 = -\frac{1}{\varrho}\frac{\partial p}{\partial s} + v\frac{\partial^2 u}{\partial n^2} \qquad \text{in the immediate proximity of the boundary} \qquad (5.18)$$

Now let us consider the velocity u along the normal direction at increasing distance from the boundary. When transitioning from the middle part of the boundary layer to the inviscid flow, u grows, approaching U. However, this increase is decreasing towards the outside, i.e. we have

$$\frac{\partial^2 u}{\partial n^2} < 0 \qquad \text{when transitioning to inviscid flow} \qquad (5.19)$$

As shown in Fig. 5.5, the flow is accelerated until the velocity reaches a maximum, while the pressure decreases. Subsequently, the velocity decreases. Now the pressure increases along the flow, i.e. in the boundary region of deceleration we find that

$$\frac{\partial p}{\partial s} > 0 \qquad\qquad (5.20)$$

and by (5.18)

$$\frac{\partial^2 u}{\partial n^2} > 0 \qquad \text{in the immediate proximity of the boundary} \qquad (5.21)$$

From (5.19) and (5.21) it follows that $\partial^2 u / \partial n^2$ changes its sign along a normal in the region of deceleration. Consequently, we find an inflection point.

> By $u(n)$ we denote the velocity as a function of its n-coordinate, whereby the s-coordinate is constant.

In Fig. 5.6, we plot $u(n)$ perpendicular to the corresponding normal and obtain the *flow profiles*. At a certain stage, they show an inflection point. When changing from a concave to a convex shape, a counterflow can occur near the boundary, causing a flow separation. This counterflow can either completely separate or reattach.

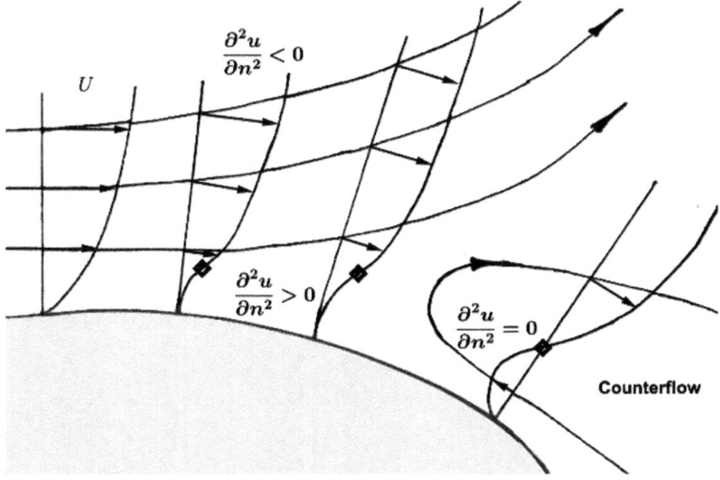

Fig. 5.6 Flow profiles after passing the velocity maximum. The formation of an inflection point (*open diamond*) can trigger a counterflow.

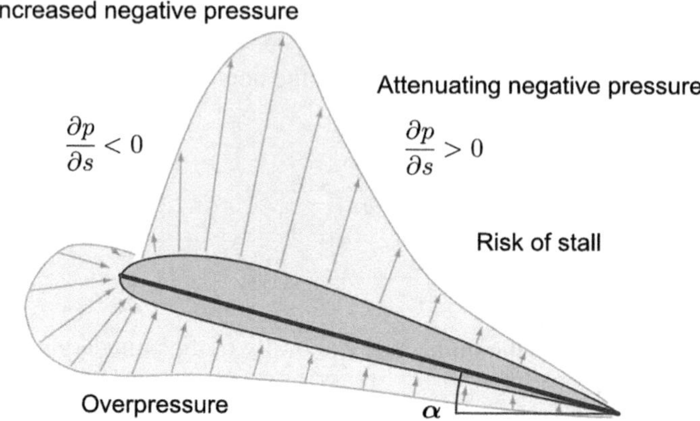

Fig. 5.7 Pressure on the airfoil profile. The arrows pointing towards the profile symbolise a pressure above the atmospheric normal pressure (overpressure), while arrows pointing away from the profile indicate a negative pressure. On the right upper side, $\partial p/\partial s > 0$, which enables a stall. Authors: C. N. Eastlake, C. R. Nave.

The same principle applies to airfoils. Since a separation on the upper side must be strictly avoided, their function is determined by two basic rules.

- The profile should taper off flat and pointed. This delays separation towards the wing tip, while rounded tail shapes promote it earlier.
- The angle between the inflow and the wing profile axis, the so-called *angle of attack* α (Fig. 5.7), should not exceed a critical value. At steeper angles of attack, the particles in the boundary flow are decelerated more strongly after passing the nose, which increases the pressure rise $\partial p/\partial s$ and the risk of flow separation on the upper side. This phenomenon is known as a *stall*.

The behaviour of boundary flows also plays an important role in locomotion on land and in water.

Due to the flow separation, vorticity can penetrate from the boundary layer into larger, surrounding areas. These zones, known as *eddy water*, increase the drag of moving objects.

Fig. 5.8 Gentoo penguin (Pygoscelis papua). The streamlined shape largely reduces energy losses.

A streamlined shape reduces eddy water and minimizes the corresponding energy losses (see Fig. 5.8).

5.5 Estimation of the Thickness of a Boundary Layer

The angle of attack at which an airfoil causes a stall decreases with lower Reynolds numbers. This reduces flight stability.

The problem has gained importance with the development of drones, especially Micro Air Vehicles ($L \leq 10$ cm, $U \approx 10$ m/s, $1000 \leq \text{Re} \leq 100{,}000$) and can be explained by the boundary layer thickness (see [6] and Example 5.2).

In the following, we use the Blasius method to solve the boundary layer Eqs. (5.12), (5.13), (5.14) for steady flow on a horizontal, infinitely thin plate. This allows us to determine the thickness $\delta(x)$ of the boundary layer. Under the assumptions made, the pressure within the boundary layer is approximately constant, i.e. the pressure gradient disappears. This simplifies the boundary layer equations.

$$u\frac{\partial u}{\partial x} + v\frac{\partial u}{\partial y} = v\frac{\partial^2 u}{\partial y^2} \tag{5.22}$$

$$\frac{\partial u}{\partial x} + \frac{\partial v}{\partial y} = 0 \tag{5.23}$$

$$u = v = 0 \qquad \text{for } y = 0 \tag{5.24}$$

Using the stream function ψ, we find that

$$\frac{\partial \psi}{\partial x} = -v, \qquad \frac{\partial \psi}{\partial y} = u \tag{5.25}$$

From (5.22) it follows that

$$\frac{\partial \psi}{\partial y}\frac{\partial^2 \psi}{\partial xy} - \frac{\partial \psi}{\partial x}\frac{\partial^2 \psi}{\partial y^2} = \nu \frac{\partial^3 \psi}{\partial y^3} \tag{5.26}$$

We determine the boundary conditions.

- $y = 0$ implies $v = 0$. Without loss of generality, we get

$$\psi = 0 \qquad \text{for } y = 0 \tag{5.27}$$

- $y = 0$ implies $u = 0$. Thus,

$$\frac{\partial \psi}{\partial y} = 0 \qquad \text{for } y = 0 \tag{5.28}$$

- $y \to \infty$ implies $u \to u_\infty$. Therefore,

$$\frac{\partial \psi}{\partial y} \to u_\infty \qquad \text{for } y \to \infty \tag{5.29}$$

The partial differential Eq. (5.26) can be transformed into an ordinary differential equation using a similarity transformation. We search for a new variable η with the approach

$$\eta = \frac{Ay}{x^k}, \tag{5.30}$$

where the constants A, k are determined later. Then we have

$$\frac{\partial \eta}{\partial x} = -\frac{k\,Ay}{x^{k+1}} = -\frac{k\eta}{x}, \qquad \frac{\partial \eta}{\partial y} = \frac{A}{x^k} \tag{5.31}$$

We introduce another constant B and a function $f(\eta)$ using the approach

$$\psi = Bx^k f(\eta) \tag{5.32}$$

Then it follows that

$$u = \frac{\partial \psi}{\partial y} = Bx^k \frac{\mathrm{d}f}{\mathrm{d}\eta}\frac{\partial \eta}{\partial y} = AB\frac{\mathrm{d}f}{\mathrm{d}\eta} \tag{5.33}$$

Now we determine the constants. If we set $AB = u_\infty$, we get

$$u = u_\infty \frac{\mathrm{d}f}{\mathrm{d}\eta} \tag{5.34}$$

Furthermore, we have

$$v = -\frac{\partial \psi}{\partial x} = -\frac{k u_\infty x^{k-1}}{A}\left(\eta \frac{\mathrm{d}f}{\mathrm{d}\eta} - f\right) \tag{5.35}$$

$$\frac{\partial^2 \psi}{\partial y^2} = \frac{\partial u}{\partial y} = \frac{u_\infty A}{x^k}\frac{\mathrm{d}^2 f}{\mathrm{d}\eta^2}$$

$$\frac{\partial^2 \psi}{\partial xy} = \frac{\partial u}{\partial x} = -\frac{k u_\infty A y}{x^{k+1}}\frac{\mathrm{d}^2 f}{\mathrm{d}\eta^2}$$

$$\frac{\partial^3 \psi}{\partial y^3} = \frac{\partial^2 u}{\partial y^2} = \frac{u_\infty A^2}{x^{2k}}\frac{\mathrm{d}^3 f}{\mathrm{d}\eta^3}$$

Now we substitute the last expressions into (5.26). After summarizing we get

$$k u_\infty^2 f \frac{\mathrm{d}^2 f}{\mathrm{d}\eta^2} = -\frac{\nu u_\infty A^2}{x^{2k-1}}\frac{\mathrm{d}^3 f}{\mathrm{d}\eta^3}$$

For

$$k = \frac{1}{2} \tag{5.36}$$

the power of x on the right side disappears and we get the ordinary differential equation

$$\frac{u_\infty}{2\nu A^2} f \frac{\mathrm{d}^2 f}{\mathrm{d}\eta^2} + \frac{\mathrm{d}^3 f}{\mathrm{d}\eta^3} = 0$$

If we choose our third constant as

$$A = (\frac{u_\infty}{\nu})^{1/2}, \tag{5.37}$$

we get

$$\frac{f}{2}\frac{\mathrm{d}^2 f}{\mathrm{d}\eta^2} + \frac{\mathrm{d}^3 f}{\mathrm{d}\eta^3} = 0 \tag{5.38}$$

Using (5.30), the boundary conditions are determined as follows.

- From (5.27) and (5.32):

$$f(0) = 0 \tag{5.39}$$

- From (5.28) and (5.33):

$$f'(0) = 0 \tag{5.40}$$

- From (5.29) and (5.33):

$$\lim_{\eta \to \infty} f'(\eta) = 1 \tag{5.41}$$

Equation (5.38) with the initial values (5.39), (5.40), (5.41) can be solved numerically applying a shooting method based on the Runge-Kutta method. The values are given in Sect. A.11, Table A.1.

Now we can estimate the thickness of the boundary layer. Summarizing (5.30), (5.36) and (5.37), we find that

$$\eta = \left(\frac{u_\infty}{\nu}\right)^{1/2} \frac{y}{x^{1/2}} \tag{5.42}$$

The value $y = y(x)$ satisfying

$$u(x, y) = 0.99\, u_\infty$$

is called the *boundary layer thickness* $\delta(x)$.

Using (5.34) we obtain

$$0.99\, u_\infty = u_\infty f'(\eta)$$

$$0.99 = f'(\eta)$$

From Table A.1 we get $\eta \approx 5$. The boundary layer thickness increases with increasing distance x from the tip. Accordingly, the Reynolds number is not constant. The value of x can be considered as a characteristic length for the definition of the corresponding Reynolds number Re_x and it holds

$$\mathrm{Re}_x = \frac{u_\infty x}{\nu}$$

Taking into account $y = \delta(x)$ and $\eta \approx 5$, we obtain the boundary layer thickness by inserting $\nu = u_\infty x / \mathrm{Re}_x$ into (5.42).

$$\delta(x) = \frac{5x}{\sqrt{\mathrm{Re}_x}} \tag{5.43}$$

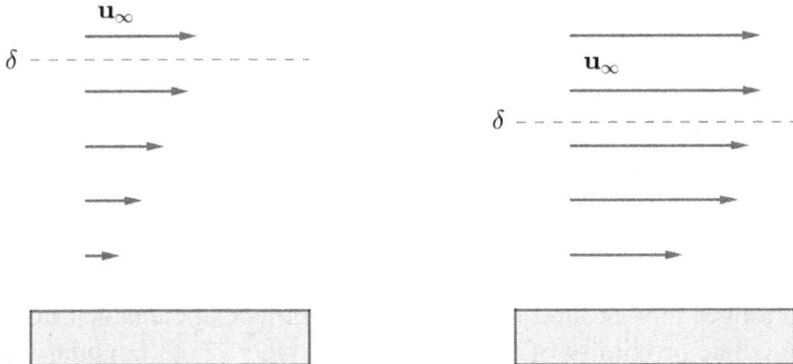

Fig. 5.9 (left) Low Reynolds number. (right) Higher Reynolds number.

Example 5.1 Small, slow aircraft have Reynolds numbers Re $\approx 2 \cdot 10^6$. At a distance of $x = 1$ m from the leading edge of the wing, the boundary layer thickness is

$$\delta = \frac{5}{\sqrt{2} \cdot 10^3} = 3.5\,\text{mm}$$

Within this narrow range, the horizontal speed u increases from zero to 99% of the inflow speed u_∞. Strong friction leads to vortex formation. Although the thickness of the boundary layer appears small, it should not be neglected. Joukowski's two-dimensional model solves this problem by incorporating the boundary layer effect through the circulation immediately adjacent to the boundary.

Example 5.2 Considering (5.43), it becomes clear why wings at higher Reynolds numbers have a lower risk of early flow separation. With a smaller δ, the flow reaches a high velocity at smaller boundary distances, thus preventing counterflows (Fig. 5.9).

5.6 The Principle of Lift Generation

So far, we have only assumed the existence of a circulation. Now we explain under which conditions our vortex field generates the circulation required for lift. We assume a slender, pointed wing profile that allows for flow separation at the wing tip according to Sect. 5.4.

First, we examine the profile at rest. Since $\mathbf{u} = 0$ applies throughout the area, the field is vortex-free at the start time, i.e. rot $\mathbf{u} = 0$. The circulation (4.10) over closed curves also disappears.

If we gradually set the profile in motion, a flow with two stagnation points similar to an ellipse is generated (see Sect. 4.6). The position of the second stagnation point results from the lower flow speed of the lower air stream compared to the upper one. The air accumulates below the wing tip and finally deflects upwards. The lower air stream rounds the wing tip, which explains the position of S_2 on the upper side of the profile (Fig. 5.10).

The deceleration of fluid particles in the profile's boundary layer creates vorticity which is bound to particles and transported with them towards the tail. As the inflow velocity increases, the frictional forces increase and turbulence is created, leading to the dissolution of the upper stagnation point. Now the particles below the wing tip can no longer flow around it. The chosen profile shape enables flow separation in the immediate proximity of the trailing edge, where both boundary flows merge at different velocities. The generated vorticity is transported downstream in a curve, where the velocity field is discontinuous (Fig. 5.11 and Lemmata 3.5, 3.6). This curve is called a *vortex sheet*. (Recall from Fig. 1.1 that a curve in the plane corresponds to a surface in space.)

Fig. 5.10 At lower flow velocities before takeoff, the air flows from below over the wing tip. Two stagnation points are formed on the profile.

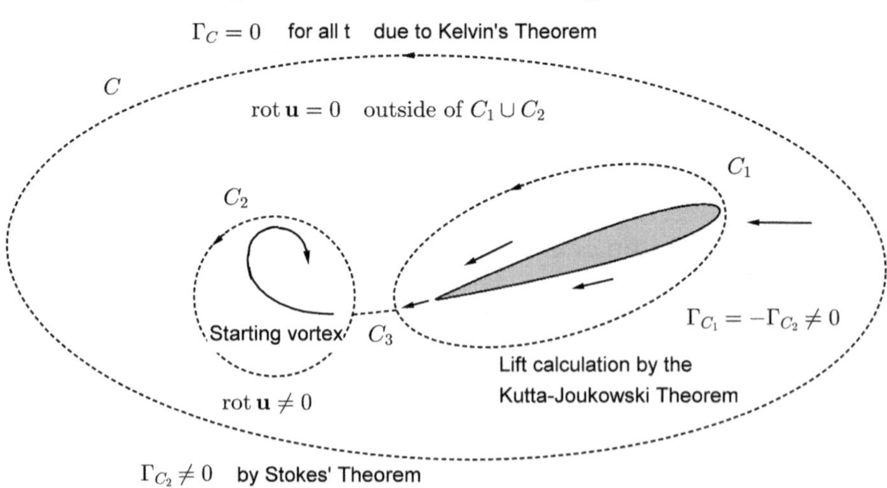

Fig. 5.11 Lift generation through friction and vortex formation. The circulation is calculated along the *dashed curves*, while *solid arrows* illustrate the flow velocities.

This layer rolls up and generates a rotating flow, which is called a *starting vortex*. By Stokes' theorem (Theorem 1.2), this corresponds to a non-vanishing circulation

$$\Gamma_{C_2} = \int_{C_2} \mathbf{u}\, d\mathbf{r} = \int \mathbf{n}\cdot \mathrm{rot}\,\mathbf{u}\, dS \neq 0$$

The flow region contains neither sources nor sinks and the vortex formation caused by flow separation is limited to the vicinity of the trailing edge, the inner region of C_2 and the narrow connecting strip C_3. Approximately, we can consider the airspace outside $C_1 \cup C_2$ as vortex-free. When the steady state is reached, the circulation around the profile remains constant.

In Fig. 5.11 we consider the annular area G between C and $C_1 \cup C_3 \cup C_2$, where the outer boundary curve C is chosen at a sufficient distance from the wing. According to Kelvin's theorem (Theorem 5.1), the initial circulation $\Gamma_C = 0$ remains unchanged. Applying Stokes' theorem, we obtain

$$0 = \int_G \mathbf{n}\cdot \mathrm{rot}\,\mathbf{u}\, dS = \int_{C\cup C_1 \cup C_3 \cup C_2} \mathbf{u}\, d\mathbf{r} \approx \Gamma_C + \Gamma_{C_1} + \Gamma_{C_2} = \Gamma_{C_1} + \Gamma_{C_2}$$

Hence, $\Gamma_{C_1} = -\Gamma_{C_2}$. *We have demonstrated a circulation around the wing that is directed in the opposite direction to the initial vortex.*

In fact, the vortex-generating boundary layer is very thin. This allows the area around the wing to be modeled using the (inviscid) Euler equations, which can be further simplified due to the lack of vorticity. Owing to $\Gamma_{C_2} \neq 0$, we obtain from the Kutta-Joukowski formula (4.13) a lift that acts perpendicular to the flow velocity.

Within the framework of our theory, we can even calculate the temporal increase of this force. In Sect. 5.7 we present a model proposed by Wagner [10].

A sufficient generation of lift thus depends on the following conditions:

- The inflow velocity and wing area are sufficiently large to generate the required vorticity.
- The boundary flow leaves the wing only at the wing tip, i.e. there is no risk of early flow separation (see Fig. 5.12 and Sect. 5.4).

Problem 5.3 The Kutta-Joukowski flow is based on the potential (4.65). Can we consider this approach to be vortex-free?

Fig. 5.12 Early flow separation on the upper side of a wing profile. The main part of the generated vorticity is released above the wing, which greatly reduces the lift and can lead to a crash.

5.7   Time Course of Lift Generation

The purpose of sharply tapering wing profiles is to generate a starting vortex for lift by concentrating the outgoing vorticity in a narrow strip. Since most of the surrounding flow field remains vortex-free, we can apply potential theory. The approach was used by Wagner [10] in a two-dimensional model, which estimates the lift growth from the time the full inflow velocity is reached. In Sect. 5.6, we explained that the vorticity flows in a vortex sheet. We are looking for a suitable potential.

Assumptions

For simplicity, the wing is assumed to be a thin plate at whose trailing edge the flow separates (see Fig. 5.13). We make the following assumptions:

W1 The freestream is started impulsively.
   This immediately ensures a constant inflow $\mathbf{u}_\infty$, which should be maintained throughout the entire process.

W2 The angle of attack α is very small.
   Since the generated vorticity is shed tangentially from the trailing edge (see Theorem 4.8), it maintains this direction and flows approximately with the speed u_∞ in the direction of the positive x-axis.

W3 The units are chosen so that the plate has a length of 2 and the inflow speed u_∞ has the value 1.

Assumption W1 can be realized via a short initial phase with high acceleration. In practice, this is possible when a wing hits a gust or is initially exposed to horizontal airflow and suddenly flipped by the angle α (Fig. 5.14).

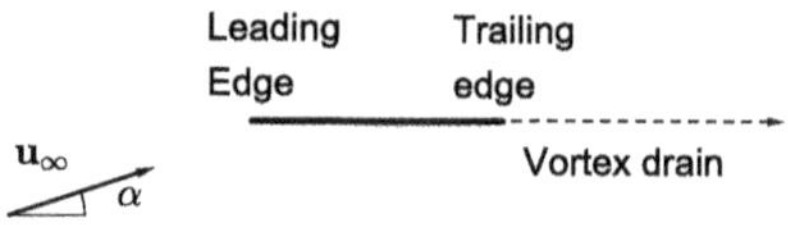

Fig. 5.13 Inflow of a plate with almost horizontal flow separation at the trailing edge. The vorticity flows out through a vortex sheet.

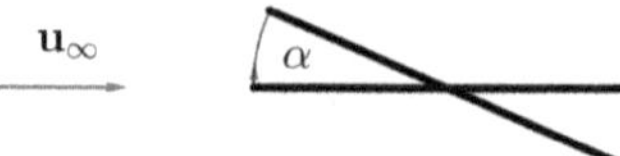

Fig. 5.14 Rapid rotation of a horizontal plate.

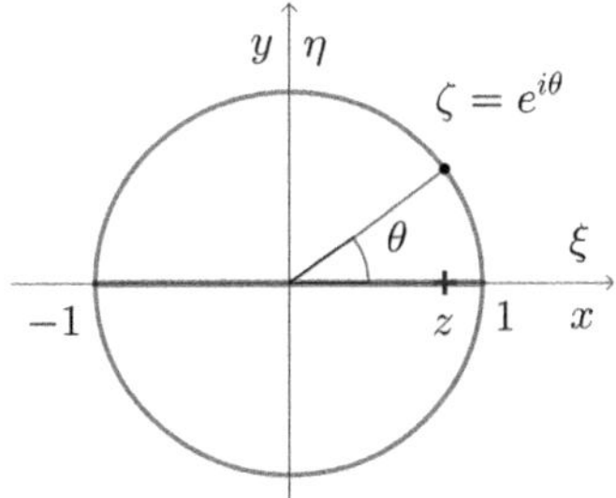

Fig. 5.15 Overlay of ζ- and z-plane. Assignment for the construction of a conformal mapping.

Transformation into the Unit Circle

The calculations are carried out by transforming the area outside the plate into the corresponding area outside a circle. As in the original paper [10] we choose a plate of length 2 to be mapped conformally into the unit circle. To determine the transformation, the boundary points

$$\zeta = e^{i\phi} \tag{5.44}$$

$$z = \cos\phi \tag{5.45}$$

are mapped onto each other (see Fig. 5.15). In analogy with Problem 4.21 it can be shown that the transformation $z = J(\zeta)$ given by

$$J(\zeta) = \frac{1}{2}\left(\zeta + \frac{1}{\zeta}\right) \tag{5.46}$$

maps the exterior regions of the plate and circle conformally onto each other. [2]

Problem 5.4   Prove

$$\frac{\zeta + 1}{\zeta - 1} = \sqrt{\frac{z + 1}{z - 1}} \tag{5.47}$$

Problem 5.5   Use Fig. 5.16 to prove the transformation rules for the plate surface

$$\zeta^{-1} - \zeta = 2\sqrt{z^2 - 1} \qquad \text{(top side)} \tag{5.48}$$

$$\zeta^{-1} - \zeta = -2\sqrt{z^2 - 1} \qquad \text{(bottom side)} \tag{5.49}$$

---

[2] Equation (5.46) differs from (4.48) by the factor $1/2$. This arises because we transform the plate into a circle of the *same* diameter. As a result of the factor, the circulation required to maintain the Joukowski condition changes.

Fig. 5.16 Overlay of ζ- and z-plane. Top and bottom of the plate are mapped onto corresponding semicircles.

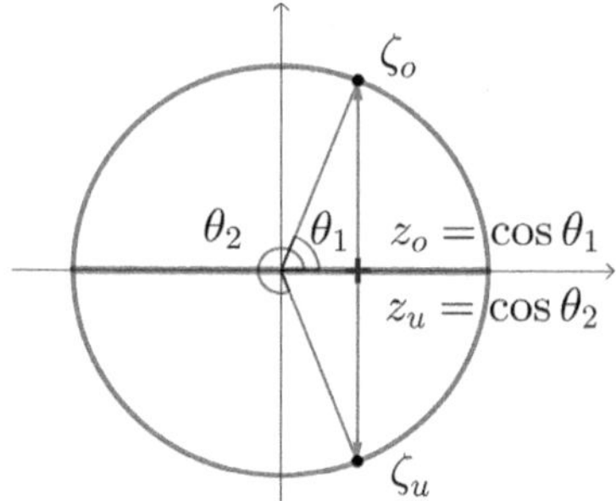

We denote the potential around the plate by $\hat{w} = \hat{\phi} + i\hat{\psi}$ and transform it conformally into the potential around the circle $w = \phi + i\psi$.

For the velocities we use the notations

$$\mathcal{U}(\zeta) \qquad \text{in the field around the circle}$$
$$\hat{\mathcal{U}}(z) \qquad \text{in the field around the plate}$$

On account of (4.17), (4.18), it holds

$$\hat{w}(z) = w(J^{-1}(z)), \qquad \mathcal{U}(\zeta) = \hat{\mathcal{U}}(J(\zeta))\,\frac{\mathrm{d}J}{\mathrm{d}\zeta}, \qquad \frac{\mathrm{d}z}{\mathrm{d}t} = \frac{\mathrm{d}J}{\mathrm{d}\zeta}\frac{\mathrm{d}\zeta}{\mathrm{d}t}$$

Let us examine how the vorticity changes when the plate is transformed onto the unit circle. According to W2, the vortex sheet generated on the plate is completely enclosed within the abscissa. In Sect. 5.6 we could see that it forms a discontinuity of the velocity field because flows from the top and bottom of the plate converge at different velocities at the trailing edge. Now we want to show that the singularity from the plate flow transforms into a singularity of the circular flow while maintaining its circulation.

Theorem 5.3 *Consider a closed curve $C \subset \mathbb{C}$ outside the unit circle under the mapping $z = J(\zeta)$ given by (5.46). Then it holds*

$$\int_C \mathcal{U}(\zeta)\,\mathrm{d}\zeta = \int_{J(C)} \hat{\mathcal{U}}(z)\,\mathrm{d}z,$$

i.e. the circulation is preserved under J.

Proof Let $C = \{\zeta(t) \mid a \le t \le b\} \subset \mathbb{C}$ be a closed curve and $J(C)$ be its image under the mapping. Then,

$$
\int_{J(C)} \hat{\mathcal{U}}(z)\,\mathrm{d}z = \int_a^b \hat{\mathcal{U}}(z(t))\,\frac{\mathrm{d}z}{\mathrm{d}t}\,\mathrm{d}t = \int_a^b \hat{\mathcal{U}}(J(\zeta(t)))\,\frac{\mathrm{d}J}{\mathrm{d}\zeta}\,\frac{\mathrm{d}\zeta}{\mathrm{d}t}\,\mathrm{d}t
$$

$$
= \int_a^b \mathcal{U}(\zeta(t))\,\frac{\mathrm{d}\zeta}{\mathrm{d}t}\,\mathrm{d}t = \int_C \mathcal{U}(\zeta)\,\mathrm{d}\zeta
$$

$\square$

Problem 5.6 Prove that the vortex density remains constant under a conformal mapping, i.e. it holds

$$
\omega(\zeta)\,\mathrm{d}\zeta = \hat{\omega}(z)\,\mathrm{d}z \quad \text{on each infinitesimal segment of the vortex sheet} \tag{5.50}
$$

The Stationary State

In contrast to Sect. 4.7, the inflow comes from the left, which requires a modification of the potential. In addition, the coordinate transformation here contains the factor $1/2$. In analogy to (4.65)–(4.67), we use the stationary circular potential

$$
w_0(\zeta) = u_\infty\left[\zeta e^{-i\alpha} + \frac{e^{i\alpha}}{\zeta}\right] - \frac{i\Gamma_0}{2\pi}\ln\zeta, \quad |\zeta| \ge 1 \tag{5.51}
$$

$$
\Gamma_0 = -2\pi u_\infty \sin\alpha \tag{5.52}
$$

as a starting point. Considering W2, it simplifies to

$$
w_0(\zeta) = u_\infty\left[\zeta + \frac{1}{\zeta}\right] - \frac{i\Gamma_0}{2\pi}\ln\zeta, \quad |\zeta| \ge 1 \tag{5.53}
$$

$$
\Gamma_0 = -2\pi u_\infty \alpha \tag{5.54}
$$

The corresponding velocity is

$$
\mathcal{U}_0(\zeta) = \frac{\mathrm{d}w_0}{\mathrm{d}\zeta} = u_\infty\left[1 - \frac{1}{\zeta^2}\right] - \frac{i\Gamma_0}{2\pi\zeta}
$$

This gives us the *detachment velocity* at the trailing edge $\zeta = 1$

$$
\mathcal{U}_0(1) = -\frac{i\Gamma_0}{2\pi} \tag{5.55}
$$

Fig. 5.17 The lift acts perpendicular to the flow and thus not perpendicular to the plate, although it results as the sum of pressure forces acting perpendicular to the surface.

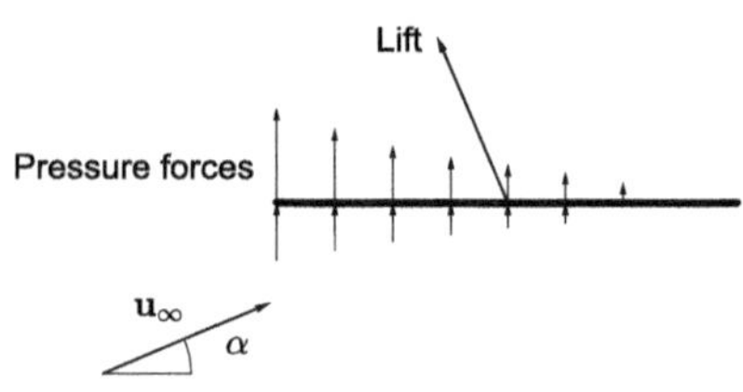

In analogy to (4.60), we obtain the velocity in z-coordinates

$$\hat{\mathcal{U}}(z) = \frac{2\mathcal{U}(\zeta)\,\zeta^2}{(\zeta - 1)(\zeta + 1)}$$

Owing to the simple geometry, the lift is easier to calculate than for the Joukowski profile. On the other hand, an infinitely thin plate leads to some unrealistic results. In contrast to the Joukowski profile, the plate has a sharp leading edge at $\zeta = -1$, where we get the velocity

$$\hat{\mathcal{U}}(-1) = \infty$$

While the singularity at the trailing edge can be avoided by the Joukowski condition (Sect. 4.7), it cannot be removed here. Furthermore, the lift acts perpendicular to the flow and and therefore not in the same direction as its components, which result from the pressure on the plate (Fig. 5.17). The paradox is caused by the strong suction force at the leading edge, the so-called *leading edge suction*, which is not taken into account. It can be avoided if we consider the plate at the leading edge as a limit case of a semicircular rounding (see [4]). According to (4.13), the stationary lift is

$$L_0 = -\varrho\Gamma_0 u_\infty \tag{5.56}$$

Decomposition of the Unsteady Vortex Field

During the start, vorticity begins to detach at the trailing edge. Then it moves in the direction of the positive x-axis and thus forms the vortex sheet. We call it *free vorticity*, in contrast to the *bound vorticity* on the plate.

Let $\omega(\zeta, t)$ be the *density of the free vorticity* for the circular potential at the location ζ at time t. Owing to (5.50), ω remains unchanged when transformed to the plate potential, i.e. it is

$$\hat{\omega}(z, t)\,dz = \omega(\zeta, t)\,d\zeta \tag{5.57}$$

The permanent shedding of vorticity at the trailing edge leads to a growing vortex sheet along the abscissa.

For the plate potential at time t, we denote the *length of the vortex sheet* by $x_{\max}(t)$. Thus, the total *amount of free vorticity* at time t is

$$\Gamma_f(t) = \int_1^{x_{\max}} \hat{\omega}(z, t)\, dz, \qquad (5.58)$$

where we integrate along the x-axis.

According to Sect. 5.6, the vanishing circulation on a widely encompassing curve enforces the opposite circulation on the wing to the vortex sheet. If we denote the *amount of bound vorticity* by $\Gamma_b(t)$, then

$$\Gamma_b(t) + \Gamma_f(t) = 0$$

Wagner's approach decomposes the entirety of vortices into pairs, assigning to the *free vortex element* $\omega(\zeta_0)\, d\zeta_0$ at ζ_0 a *bound vortex element* of opposite vortex strength $-\,\omega(\zeta_0)\, d\zeta_0$ at ζ_0^{-1} (see Fig. 5.18).

Based on the generation and transport of free vorticity, we can now calculate the increase in bound vorticity and thus the temporal development of the lift.

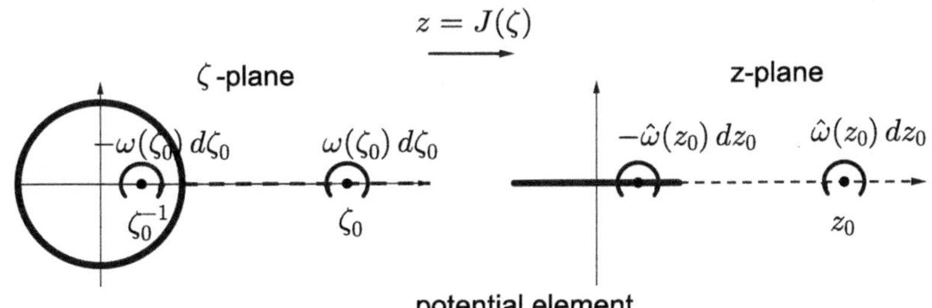

Fig. 5.18 A bound vortex element of opposite strength at ζ_0^{-1} is assigned to the free vortex element at ζ_0. According to (5.50), vorticity remains invariant under transformation into the z-plane.

A *vortex pair* with the free element at ζ_0 (in ζ- coordinates) or at z_0 (in z-coordinates), called a *vortex pair at ζ_0 or z_0* for short, refers to the pair consisting of the free vortex element at ζ_0 and the bound vortex element of opposite strength at ζ_0^{-1}.

To the vortex pair at ζ_0 we assign the potential

$$w_{\zeta_0} = u_\infty\left[\zeta + \frac{1}{\zeta}\right] + dw_{\zeta_0}(\zeta) \tag{5.59}$$

Here,

$$dw_{\zeta_0}(\zeta) = \frac{\mathrm{i}\,\omega(\zeta_0)\,d\zeta_0}{2\pi}[\ln(\zeta - \zeta_0) - \ln(\zeta - \zeta_0^{-1})] \tag{5.60}$$

denotes the *potential element* at ζ_0. If we write w_{ζ_0} and $dw_{\zeta_0}(\zeta)$ as the sum of velocity potential and stream function, we get

$$dw_{\zeta_0} = d\phi_{\zeta_0} + \mathrm{i}\,d\psi_{\zeta_0}$$
$$w_{\zeta_0} = \phi + d\phi_{\zeta_0} + \mathrm{i}(\psi + d\psi_{\zeta_0}) \tag{5.61}$$

and therefore

$$u_\infty\left[\zeta + \frac{1}{\zeta}\right] = \phi + \mathrm{i}\psi$$

In Sect. 5.6 we showed that $u_\infty[\zeta + 1/\zeta]$ is a potential for a flow around a circle. As will be demonstrated in Lemma 5.1, the same holds for dw_{ζ_0} because ζ_0^{-1} results from the inversion of ζ_0 on the circle.

Problem 5.7 Why are ζ_0, ζ_0^{-1} and z_0 real?

Applying the Joukowski transformation (5.46), we transfer the circular potential w_{ζ_0} to the plate potential $\hat{w}_{z_0}$.

$$\hat{w}_{z_0}(z) = w_{\zeta_0}(J^{-1}(z)), \qquad z_0 = J(\zeta_0) \tag{5.62}$$

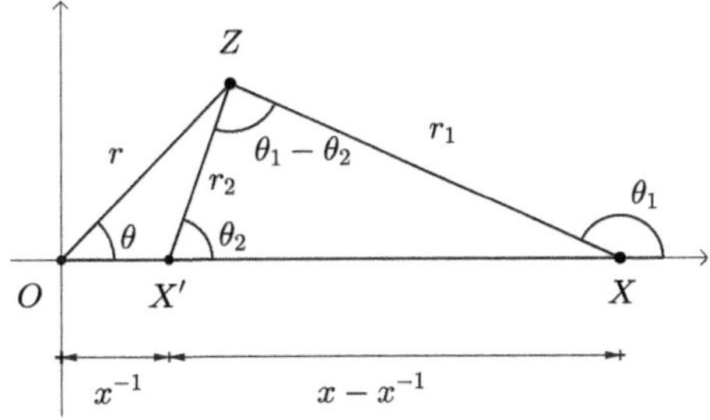

Fig. 5.19 Proof sketch for Problems 5.8–5.10.

By means of the decomposition

$$\hat{w}_{z_0} = (\hat{\phi} + d\hat{\phi}_{z_0}) + i(\hat{\psi} + d\hat{\psi}_{z_0}) \tag{5.63}$$

we obtain the corresponding velocity potential and the stream function. For a potential element, we find that

$$dw_{\zeta_0}(J^{-1}(z)) = d\hat{w}_{z_0}(z) = d\hat{\phi}_{z_0} + i\, d\hat{\psi}_{z_0}$$

Evaluation of a Potential Element

In analogy with the theorems of Blasius and Kutta-Joukowski, the time course of the lift can be calculated from the velocity of the potential elements.

Problem 5.8 In Fig. 5.19, the complex numbers ζ, ζ_0 and ζ_0^{-1} correspond to the points Z, X and X', respectively, so that

$$\zeta = re^{i\theta}, \qquad \zeta - \zeta_0 = r_1 e^{i\theta_1}, \qquad \zeta - \zeta_0^{-1} = r_2 e^{i\theta_2}$$

Prove the formula

$$dw_{\zeta_0}(\zeta) = \frac{\omega(\zeta_0)\, d\zeta_0}{2\pi}\left[(\theta_2 - \theta_1) + i\ln\frac{r_1}{r_2}\right] \tag{5.64}$$

From (5.64) we obtain the velocity potential

$$d\phi_{\zeta_0} = \frac{\omega(\zeta_0)\, d\zeta_0}{2\pi}(\theta_2 - \theta_1) \tag{5.65}$$

and the stream function

$$d\psi_{\zeta_0} = \frac{\omega(\zeta_0)\, d\zeta_0}{2\pi}\ln\frac{r_1}{r_2} \tag{5.66}$$

Problem 5.9 Show in Fig. 5.19 that

$$\left(\frac{r_1}{r_2}\right)^2 = x^2 \qquad \text{for } r = 1 \tag{5.67}$$

We check whether the potential is suitable for the flow domain.

Lemma 5.1 *The element of the stream function* $d\psi_{\zeta_0}$ *remains constant on the circle boundary* $K(0, 1)$.

Proof Owing to (5.67), the element of the stream function (5.66)

$$d\psi_{\zeta_0}(\zeta) = \frac{\omega(\zeta_0)\,d\zeta_0}{2\pi} \ln x$$

is independent of the circle point ζ. $\square$

Calculation of the Lift

For the sake of simplicity, we omit the accent mark for the flow around the plate when it is obvious, i.e. we write $\omega(z_0)$ instead of $\hat{\omega}(z_0)$ etc. (Recall that the variable z_0 refers to the flow around the plate.)

Since different flows prevail on both sides of the plate, different elements of the velocity potential are to be expected there.

We denote the *element of the velocity potential*
- on the upper side by $d\phi_{z_0}^{+}$,
- on the lower side by $d\phi_{z_0}^{-}$.

They are useful for determining the fraction by which a potential element dw_{z_0} contributes to the lift. Again, we need some trigonometry.

Problem 5.10 Prove in Fig. 5.19 that

$$\tan(\theta_1 - \theta_2) = \frac{(x - x^{-1})\sin\theta}{2 - (x + x^{-1})\cos\theta} \qquad \text{for } r = 1 \tag{5.68}$$

Lemma 5.2 *Let* $dw_{z_0} = d\phi_{z_0} + i\,d\psi_{z_0}$ *be a potential element. Then its velocity potential satisfies the following relations.*

On the upper side of the plate

$$d\phi_{z_0}^{+}(z) = \frac{\omega(z_0)\,dz_0}{2\pi} \, arctan2\left[\sqrt{z_0^2 - 1}\,\sqrt{1 - z^2},\,(1 - z_0 z)\right] \tag{5.69}$$

On the lower side of the plate

$$d\phi_{z_0}^-(z) = \frac{\omega(z_0)\,dz_0}{2\pi}\,arctan2\left[-\sqrt{z_0^2 - 1}\,\sqrt{1 - z^2},\;(1 - z_0 z)\right] \tag{5.70}$$

Proof For the definition of arctan2, we refer to Sect. A.4, formulas (A.19), (A.20). Since ζ_0, ζ_0^{-1} are real, it follows from (5.68) that

$$\theta_2 - \theta_1 = arctan2\left[(\zeta_0^{-1} - \zeta_0)\sin\theta,\;2 - (\zeta_0 + \zeta_0^{-1})\cos\theta\right]$$

and we obtain from (5.65) on the unit circle

$$d\phi_{\zeta_0} = \frac{\omega(\zeta_0)\,d\zeta_0}{2\pi}\,arctan2\left[(\zeta_0^{-1} - \zeta_0)\sin\theta,\;2 - (\zeta_0 + \zeta_0^{-1})\cos\theta\right]$$

The plate potentials are calculated by transformation into the z-plane. Directly on the plate, by (5.45),

$$\cos\theta = z, \quad \text{and therefore } \sin\theta = \sqrt{1 - z^2}$$

Taking into account (5.46), (5.48), (5.49) as well as (5.50), we obtain

$$d\phi_{z_0}^-(z) = \frac{\omega(z_0)\,dz_0}{2\pi}\,arctan2\left[-2\sqrt{z_0^2 - 1}\,\sqrt{1 - z^2},\;2(1 - z_0 z)\right]$$

$$d\phi_{z_0}^+(z) = \frac{\omega(z_0)\,dz_0}{2\pi}\,arctan2\left[2\sqrt{z_0^2 - 1}\,\sqrt{1 - z^2},\;2(1 - z_0 z)\right]$$

Considering (A.20), the factor 2 can be shortened and it follows (5.69), (5.70). $\square$

Problem 5.11 Prove

$$d\phi_{z_0}^-(z) = -d\phi_{z_0}^+(z), \qquad -1 \le z \le 1 \tag{5.71}$$

By

$$\Delta p_{z_0}(z) = p_{z_0}^+ - p_{z_0}^-, \qquad -1 \le z \le 1 \tag{5.72}$$

we denote the pressure difference at z between both sides of the plate, which is caused by potential (5.62) .

For pressure calculation, we use the unsteady Bernoulli equation (4.8), which we note for the upper and lower side. The velocity potential is taken from (5.63).

$$\frac{\partial(\hat{\phi} + d\phi_{z_0}^+)}{\partial t} + \frac{1}{2}\|\nabla(\hat{\phi} + d\phi_{z_0}^+)\|^2 + \frac{p_{z_0}^+}{\varrho} = c(t) \tag{5.73}$$

$$\frac{\partial(\hat{\phi} + d\phi_{z_0}^-)}{\partial t} + \frac{1}{2}\|\nabla(\hat{\phi} + d\phi_{z_0}^-)\|^2 + \frac{p_{z_0}^-}{\varrho} = c(t) \tag{5.74}$$

The constant $c(t)$ depends on the time, but not on the location. Since

$$\|\nabla(\hat{\phi} + d\phi_{z_0}^+)\|^2 - \|\nabla(\hat{\phi} + d\phi_{z_0}^-)\|^2$$

$$= 2\,\nabla\hat{\phi}\,\nabla(d\phi_{z_0}^+ - d\phi_{z_0}^-) + (\nabla d\phi_{z_0}^+)^2 - (\nabla d\phi_{z_0}^+)^2$$

$$\approx 2\,\nabla\hat{\phi}\,\nabla(d\phi_{z_0}^+ - d\phi_{z_0}^-),$$

we obtain after subtracting (5.74) from (5.73),

$$\frac{\partial}{\partial t}(d\phi_{z_0}^+ - d\phi_{z_0}^-) + \nabla\hat{\phi}\,\nabla(d\phi_{z_0}^+ - d\phi_{z_0}^-) + \frac{\Delta p_{z_0}}{\varrho} = 0$$

and by (5.71)

$$\Delta p_{z_0}(z) = -2\varrho\left(\frac{\partial}{\partial t}d\phi_{z_0}^+ + \nabla\hat{\phi}\,\nabla\,d\phi_{z_0}^+\right) \tag{5.75}$$

According to assumption W2, the fluid on the plate and after detachment at the trailing edge flows in the x-direction. Therefore, we find for the velocity potential $\hat{\phi}$ on the plate and the velocity of the free vortices $\mathbf{u}_f$ that

$$\nabla\hat{\phi} = \mathbf{u}_f = \begin{pmatrix} u_\infty \\ 0 \end{pmatrix} \tag{5.76}$$

It follows that

$$\nabla\hat{\phi}\,\nabla d\phi_{z_0}^+ \approx u_\infty\,\frac{\partial}{\partial x}d\phi_{z_0}^+$$

and (5.75) takes the form

$$\Delta p_{z_0}(z) = -2\varrho\left(\frac{\partial}{\partial t}d\phi_{z_0}^+ + u_\infty\,\frac{\partial}{\partial x}d\phi_{z_0}^+\right) \tag{5.77}$$

For the time derivative, we consider the velocity potential $d\phi_{z_0}$ during the movement of a free vortex. It satisfies the convection equation[3]

$$\frac{\partial}{\partial t} d\phi_{z_0}^+ = \mathrm{div}_{z_0}\,(\mathbf{u}_f\, d\phi_{z_0}^+)$$

$$= d\phi_{z_0}^+\,\mathrm{div}_{z_0}\,\mathbf{u}_f + \mathbf{u}_f\,\mathrm{grad}_{z_0}\,d\phi_{z_0}^+ \qquad \text{by (1.15)}$$

$$= u_\infty\,\frac{\partial}{\partial x_0} d\phi_{z_0}^+ \qquad \text{by (5.76)} \tag{5.78}$$

By substitution into (5.77) we obtain

$$\Delta p_{z_0}(z) = -2\varrho u_\infty \left(\frac{\partial}{\partial x_0} d\phi_{z_0}^+ + \frac{\partial}{\partial x} d\phi_{z_0}^+ \right) \tag{5.79}$$

Since both the plate and the drifting vortex are located on the x-axis, their coordinates z_0, z are real and can be replaced by x-components. In particular, we have

$$z = x \in [-1, 1] \qquad \text{for the position on the plate}$$
$$z_0 = x_0 > 0 \qquad \text{for the position of the free vortex}$$

By $l(x_0)$ we denote the *contribution to the lift*, which is *generated by the vortex pair with the free element at* x_0.

The *lift at time* $t \geq 0$ results from the sum of all contributions $l(x_0)$ and is denoted by $L(t)$.

Problem 5.12 Prove that

$$l(x_0) = -\frac{\varrho u_\infty\, \omega(x_0)\, x_0\, dx_0}{\sqrt{x_0^2 - 1}} \qquad \text{for } x_0 > 1 \tag{5.80}$$

If we start at time 0, a vortex sheet of length $x_{\max}(t)$ has been shed from the trailing edge at $t > 0$.

---

[3] See convection-diffusion equation. Another argument for understanding (5.78) is the consideration of $c(x_0) = d\phi_{z_0}^+$ for a moving free vortex. From $u_\infty = \frac{\partial x_0}{\partial t} = \frac{\partial c}{\partial t}\frac{\partial x_0}{\partial c} = \frac{\partial c}{\partial t} / \frac{\partial c}{\partial x_0}$ it follows that $\frac{\partial c}{\partial t} = u_\infty \frac{\partial c}{\partial x_0}$.

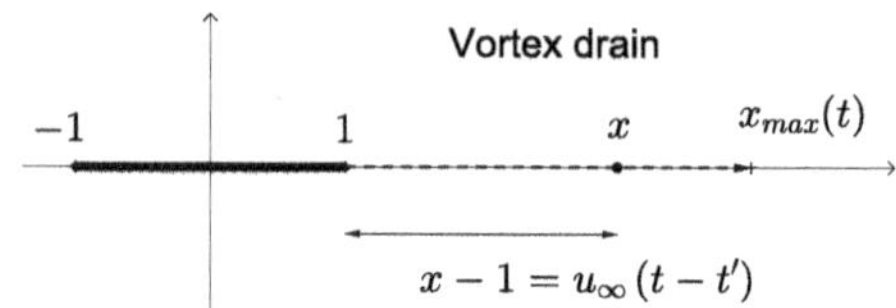

Fig. 5.20 A vortex that has detached from the trailing edge at time t' has reached the location x at time t.

By adding the contributions (5.80) of all $x_0 \in (1, x_{\max}(t)]$ and renaming $x = x_0$ we get the lift at time $t > 0$

$$L(t) = \int_1^{x_{\max}(t)} l(x)\, dx = -\varrho u_\infty \int_1^{x_{\max}(t)} \omega(x)\frac{x\, dx}{\sqrt{x^2 - 1}}$$

Now we apply a variable transformation.

For a vortex sheet at time $t > 0$

$$1 \le x \le x_{\max}(t),$$

we assign to each location x the time t' at which the vorticity $\omega(x)$ has detached at the trailing edge. By

$$\mu(t') = \omega(x)$$

we denote the density of the vortex sheet.

According to Fig. 5.20, we have

$$x = 1 + (t - t')u_\infty$$
$$x = 1 + (t - t') \qquad \text{using W3} \tag{5.81}$$
$$dx = -dt'$$

Using W3 again, we obtain for the lift at time t

$$L(t) = -\varrho \int_0^t \mu(t')\frac{1 - t - t'}{\sqrt{(t - t')(t - t' + 2)}}\, dt' \tag{5.82}$$

The ratio of unsteady to steady lift is referred to as the *Wagner function*

$$\Phi(t) = \frac{L(t)}{L_0}$$

We call

$$\tilde{\mu}(t') = \frac{\mu(t')}{L_0}$$

the *normalized wake vorticity distribution.*

From (5.82), we find that

$$\Phi(t) = -\varrho \int_0^t \tilde{\mu}(t') \frac{1 - t - t'}{\sqrt{(t - t')(t - t' + 2)}} \, \mathrm{d}t' \tag{5.83}$$

Determination of the Wagner Function

To evaluate (5.83), we need to find the normalized wake vorticity distribution $\mu(t')$. We determine it for the steady state, taking into account the detachment speed at the trailing edge (5.55). Referring to W3 we set $u_\infty = 1$.

Using (5.59) and (5.60), we see that a vortex pair at ζ_0 generates the velocity field

$$\mathcal{U}_{\zeta_0}(\zeta) = \frac{\mathrm{d}w_{\zeta_0}}{\mathrm{d}\zeta} = 1 - \zeta^{-2} + \frac{\mathrm{i}\,\omega(\zeta_0)\,\mathrm{d}\zeta_0}{2\pi} \left[\frac{1}{\zeta - \zeta_0} - \frac{1}{\zeta - \zeta_0^{-1}} \right]$$

and thus the velocity at the trailing edge

$$\mathcal{U}_{\zeta_0}(1) = \frac{\mathrm{i}\,\omega(\zeta_0)\,\mathrm{d}\zeta_0}{2\pi} \left[\frac{1}{1 - \zeta_0} - \frac{1}{1 - \zeta_0^{-1}} \right] = -\frac{\mathrm{i}\,\omega(\zeta_0)\,\mathrm{d}\zeta_0}{2\pi} \frac{\zeta_0 + 1}{\zeta_0 - 1}$$

Upon transformation into the z-plane, we get from (5.47)

$$\hat{\mathcal{U}}_{z_0}(1) = -\frac{\mathrm{i}\,\omega(z_0)\,\mathrm{d}z_0}{2\pi} \sqrt{\frac{z_0 + 1}{z_0 - 1}}$$

By summing over all vortex pairs, we obtain the velocity at the trailing edge at time t

$$\hat{\mathcal{U}}(1) = -\frac{i}{2\pi} \int_1^{x_{\max}(t)} \omega(z_0) \sqrt{\frac{z_0 + 1}{z_0 - 1}}\, dz_0$$

$$= -\frac{i}{2\pi} \int_0^t \mu(t') \sqrt{\frac{t - t' + 2}{t - t'}}\, dt' \quad \text{using transformation (5.81)}$$

For $t \to \infty$, we find the detachment velocity of the steady state (5.55).

$$-\frac{i\Gamma_0}{2\pi} = -\frac{i}{2\pi} \lim_{t \to \infty} \int_0^t \mu(t') \sqrt{\frac{t - t' + 2}{t - t'}}\, dt'$$

Owing to (5.56) and W3, it holds that $\Gamma_0 = -L_0/\varrho$ and the equation simplifies to

$$1 = \lim_{t \to \infty} \int_0^t \tilde{\mu}(t') \sqrt{\frac{t - t' + 2}{t - t'}}\, dt' \tag{5.84}$$

From the integral Eqs. (5.83) and (5.84), we can find a numerical solution for the Wagner function (see [2]) and obtain Fig. 5.21. Owing to W3, we have $u_\infty = 1$, so that half of a plate length is covered in one unit of time.

A *half-chord* denotes the time required to travel half the length of the plate.

Figure 5.21 shows that the lift is already half of the maximum value immediately after full inflow. The further build-up is delayed, so that a value of 95% is only reached after 20 half-chords.

Since the evaluation of the integral equations is challenging, a number of approximate solutions have also been proposed (see [2]).

Fig. 5.21 Wagner function.

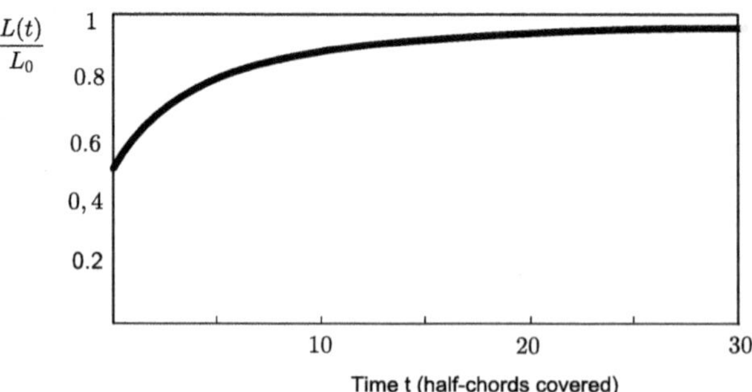

The approximation by explicit formulas allows their use in non-stationary models, for example for the analysis of insect wing beats (see [8, 9]).

5.8   The Laws of Helmholtz and Biot-Savart

In this section, we show how vorticity influences the surrounding velocity field. We denote the *vorticity* by $\boldsymbol{\omega} = \mathrm{rot}\,\mathbf{u}$ and by *magnitude* with $\omega = \|\boldsymbol{\omega}\|$.

A *vortex line* is an curve in the flow domain, that is everywhere tangential to the vorticity field.

Consider a surface S in the flow domain that is nowhere tangential to the vorticity field. A *vortex tube* is formed by all vortex lines that intersect ∂S.

A *vortex filament* is a vortex tube with an infinitesimally small cross-section, so that the fluid properties can be assumed to be constant across the cross-section (Fig. 5.22).

For the vortex tube, we assume that it is tubular in the intuitive sense and does not contain any branches, for example. The concept of the infinitesimal vortex filament is also intuitive. For a mathematical proof of the following theorem, our definition is not sufficient, see [1]. Therefore, we limit ourselves to a plausibility reasoning.

Theorem 5.4 (Helmholtz) *Consider a flow with constant density, which is determined by (3.39) and (3.40). Then, the following statements are true.*

(1) The magnitude of vorticity remains constant at each point of the vortex filament.
(2) Vortex filaments do not end in the fluid. If they do not run to the boundary, they either form closed or infinite curves.

Plausibility Check   From (1.18) we obtain

$$\mathrm{div}\,\boldsymbol{\omega} = \mathrm{div}\,\mathrm{rot}\,\mathbf{u} = 0 \tag{5.85}$$

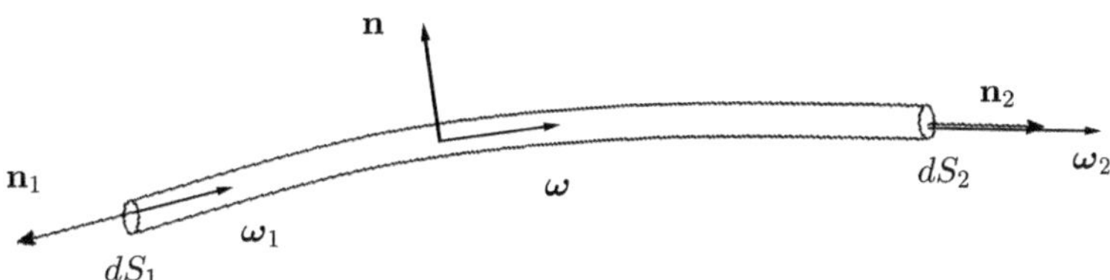

Fig. 5.22 Vortex filament.

Ad. (1):   We consider the portion of a vortex filament that is enclosed between dS_1 and dS_2 (Fig. 5.22). Applying Gauss's theorem and (5.85), we find that

$$\int_S \boldsymbol{\omega} \cdot \mathbf{n}\, dS = \int_V \operatorname{div} \boldsymbol{\omega}\, dV = 0,$$

where S denotes the surface and V the enclosed volume. The surface consists of the mantle S_m and the infinitesimal caps $dS_1 = dS_2$. Since the vorticity runs in the tube direction, it holds $\boldsymbol{\omega} \cdot \mathbf{n} = 0$ for the mantle normal $\mathbf{n}$, see Fig. 5.22. Thus, the surface integral is

$$0 = \int_S \boldsymbol{\omega} \cdot \mathbf{n}\, dS = \int_{dS_1} \boldsymbol{\omega} \cdot \mathbf{n}\, dS + \int_{dS_2} \boldsymbol{\omega} \cdot \mathbf{n}\, dS$$

Taking into account the directions in Fig. 5.22 and the definition of the vortex filament, we get

$$\int_{dS_1} \boldsymbol{\omega} \cdot \mathbf{n}\, dS = -\omega_1\, dS_1, \qquad \int_{dS_2} \boldsymbol{\omega} \cdot \mathbf{n}\, dS = \omega_2\, dS_2$$

and consequently $\omega_1 = \omega_2$.

Ad. (2):   We assume that our vortex filament does not branch, so that the vorticity does not vanish anywhere (see [1]). If the vortex filament ends at dS_2 in the fluid, it follows that $\omega_2 = 0$ and we get a contradiction to the first assertion.

Problem 5.13 A vortex tube with a sufficiently small cross-section can be considered as a vortex filament. Show that the strength ω of the vortex filament is calculated by

$$\omega = \lim_{|\Sigma| \to 0} \frac{\Gamma_C}{|\Sigma|}$$

from the circulation Γ_C of a corresponding vortex tube, where C denotes a surrounding curve, Σ the enclosed cross-section and $|\Sigma|$ the area measure for the tube cross-section (Fig. 5.23).

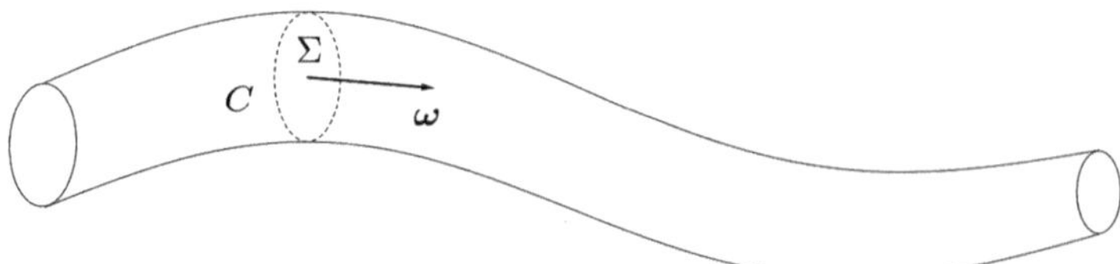

Fig. 5.23 Vortex tube with surrounding curve and cross-section.

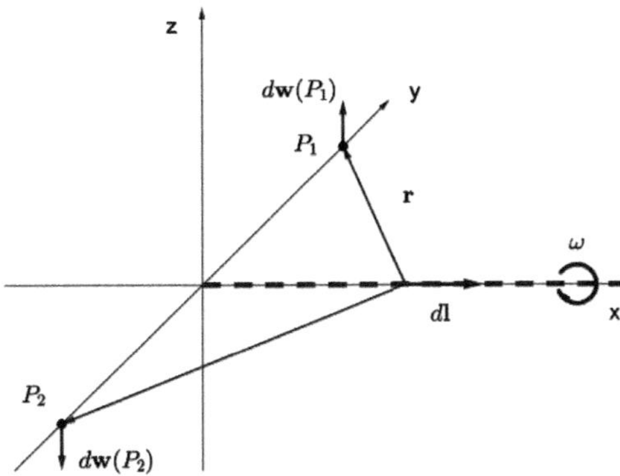

Fig. 5.24 The vortex filament (dashed) induces a velocity field.

Theorem 5.5 (Biot-Savart) *Consider a flow with constant density, which is determined by (3.39) and (3.40). Let* $\mathbf{dl}$ *be the infinitesimal segment of a vortex filament of strength ω and P a point in space. Then* $\mathbf{dl}$ *induces the flow velocity at P*

$$\mathbf{dw} = \frac{\omega}{4\pi} \frac{\mathbf{dl} \times \mathbf{r}}{\|\mathbf{r}\|^3}, \tag{5.86}$$

where $\mathbf{r}$ denotes the distance vector from $\mathbf{dl}$ *to P (Fig. 5.24).*

In addition to fluid mechanics, the law is applied in electrodynamics, where it describes the magnetic field of moving charges. Here, the current density $\mathbf{j}$ replaces the vorticity $\boldsymbol{\omega}$ and the induced magnetic field strength $\mathbf{B}$ corresponds to the induced flow velocity. An integral form of the formula is derived in Sect. A.9.

In Fig. 5.24 we consider two points P_1, P_2, which are symmetrical to the vortex filament. From (5.86) it follows that their velocities induced by $\mathbf{dl}$, $\mathbf{dw}(P_1)$ and $\mathbf{dw}(P_2)$, are directed oppositely. The different flow directions will play a crucial role for the lift of airfoils in Chap. 6.

Example 5.3 A radial vortex filament of strength ω runs along the positive x-axis, see Fig. 5.24. We calculate the induced velocity $\mathbf{w}$ at a point on the y-axis.

The vortex filament $0 \leq x < \infty$ runs in the x-direction. According to Fig. 5.25, we assign the vector

$$\mathbf{r} = \overrightarrow{QP} = -x(\phi)\,\mathbf{e}_x + y_1\,\mathbf{e}_y$$

to the segment $\mathbf{dl} = \mathbf{e}_x\,dx$ starting at Q. Hence,

$$\mathbf{dl} \times \mathbf{r} = y_1\,dx\,\mathbf{e}_z$$

Fig. 5.25 Vortex filament (*dashed*) in top view.

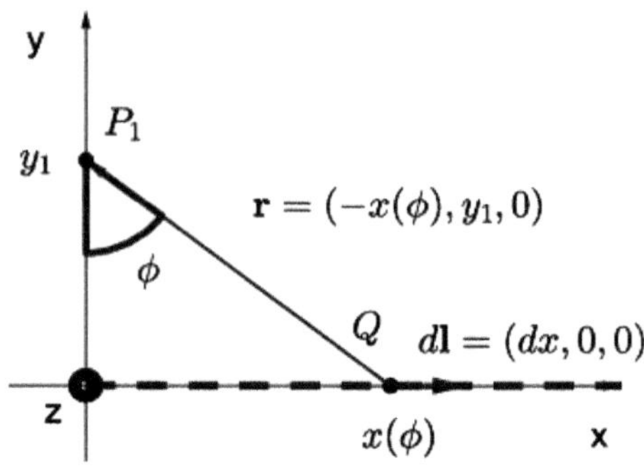

For $r = \|\mathbf{r}(\phi)\|$ it follows that

$$x(\phi) = y_1 \tan\phi, \quad \mathrm{d}x = \frac{y_1}{\cos^2\phi}\,\mathrm{d}\phi \quad \text{and} \quad \cos\phi = \frac{y_1}{r}$$

Then

$$\frac{\mathrm{d}x}{r^3} = \frac{\cos\phi}{y_1^2}\,\mathrm{d}\phi$$

and we obtain by integrating along the positive x-axis

$$\mathbf{w} = \frac{\omega}{4\pi}\int_{\mathbb{R}^+}\frac{\mathrm{d}\mathbf{l}\times\mathbf{r}}{\|\mathbf{r}\|^3} = \frac{\omega y_1\,\mathbf{e}_z}{4\pi}\int_{\mathbb{R}^+}\frac{\mathrm{d}x}{r^3} = \frac{\omega\,\mathbf{e}_z}{4\pi y_1}\int_0^{\pi/2}\cos\phi\,\mathrm{d}\phi$$
$$= \frac{\omega\,\mathbf{e}_z}{4\pi y_1}, \tag{5.87}$$

where y_1 denotes the distance from P to the radial vortex filament.

Problem 5.14 Whirlpools form in rivers due to irregularities on the bottom, such as stuck tree debris. If you are in danger of getting caught in such an eddy, experts recommend crossing it as shallowly as possible on the surface. Give reasons for this recommendation.

References and Further Reading

1. Chorin A.J., Marsden J.E. : *A Mathematical Introduction to Fluid Mechanics*. Springer, Berlin (1992)
2. Dawson S.T.M., Brunton S.L.: *Improved approximations to the Wagner function using sparse identification of nonlinear dynamics*. AIAAJ, Vol. 60, N. 3 (March 2022).
3. Goldstein S.: *Modern Developments in Fluid Dynamics*. Dover Publications, New York (1938), pp. 119–120.

 4. Katz J., Plotkin A.: *Low-Speed Aerodynamics*. Cambridge University Press, Cambridge (2001), pp. 153.
 5. M.F. Platzer, K.D. Jones: *Concepts of aero- and hydromechanics*. WIT, Flow Phenomena in Nature, Vol. 1 (2006), pp. 5–19
 6. Platzer M.F., Jones K.D.: *Steady and unsteady aerodynamics*. WIT, Flow Phenomena in Nature, Vol. 2 (2006), pp. 531–541
 7. Rosenhead L.: *Laminar Boundary Layers*. Clarendon Press, Oxford (1963), pp.201–203.
 8. Shyy W., Aono H., Kang C.K., Liu H.: *An Introduction to Flapping Wing Aerodynamics*. Cambridge University Press, Cambridge (2013)
 9. Sihite E., Ghanem P., Salagame A., Ramezani A.: *Unsteady aerodynamic modeling of Aerobat using lifting line theory and Wagner's function*. IEEE/RSJ International Conference on Intelligent Robots and Systems (IROS), 25 July 2022.
10. Wagner H.: *Über die Entstehung des dynamischen Auftriebs von Tragflügeln*. Zeitschrift für Angewandte Mathematik und Mechanik 5, N. 1 (Febr. 1925).

Lifting Line Theory

6

Now we calculate the lift on the three-dimensional wing for the steady state, i.e. already in flight. We use the two-dimensional results of Kutta-Joukowski, where the vorticity is incorporated by the trick of circulation. The latter is contained in the sectional lift coefficient, which is measured for each given wing profile in the wind tunnel. Thus, we can treat the problem based on the Euler equations. The approach was first formulated by Lanchester at the beginning of the twentieth century and further developed independently by Prandtl.

6.1 Theoretical Foundations

Figure 6.1 shows a wing embedded in a coordinate system in a specific way.

Wing Geometry

- b —*Wing Span*: the length of the wing in the y-direction (Fig. 6.1).

  The distance $-\frac{b}{2} \leq y \leq \frac{b}{2}$ is referred to as the *lifting line* (Fig. 6.1 left).

- $c(y)$ —*Wing Chord* at y, see Fig. 6.1 right.

  In particular, $c_r = c(0)$ is the *Wing Root* and $c_t = c(\pm b/2)$ is the *Wing Tip*.

- S —*Wing Area*, see Fig. 6.1 right.

- $\mathcal{R} = \frac{b^2}{S}$ —*Aspect Ratio*

- A cross-section along the y-axis is referred to as the *Profile* at y. See shaded areas in Fig. 6.1 left.

The original version of the chapter has been revised. A correction to this chapter can be found at https://doi.org/10.1007/978-3-662-72240-4_9

R. Spielmann, *Theoretical Fluid Mechanics*, https://doi.org/10.1007/978-3-662-72240-4_6

133

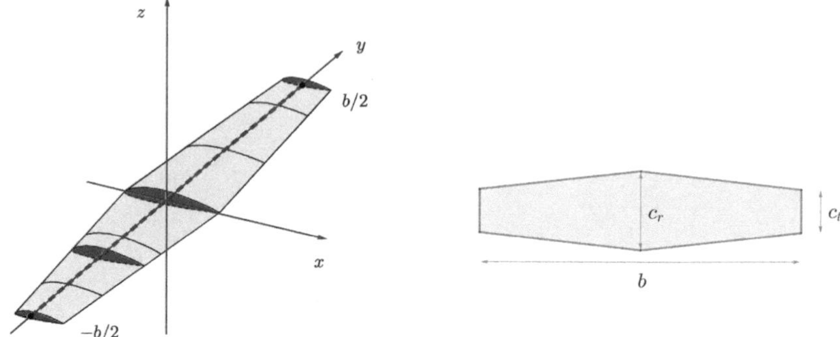

Fig. 6.1 (left) Oblique view with lifting line (dashed) (right) Top view (planform).

Fig. 6.2 The pressure equalisation over the wing edges causes a downward wind speed **w** on both sides of the wing.

Velocities, Forces and Angles

The flow over a finite wing is three-dimensional, with a considerable flow possible in the span direction. It is caused by the pressure difference between the top and bottom of the wing. The flow moves from high to low pressure, i.e. from below to the edge and upwards (Fig. 6.2).

> The *downward* wind speed **w** on the top and bottom of the wing is referred to as *downwash* (Fig. 6.2).

A vortex field is created behind the aircraft, which generates an upward flow of air on the outside and a downward flow on the inside (Fig. 6.3). Particularly strong vorticity is formed at the wing tips (see Fig. 6.4) and Problem 6.6.

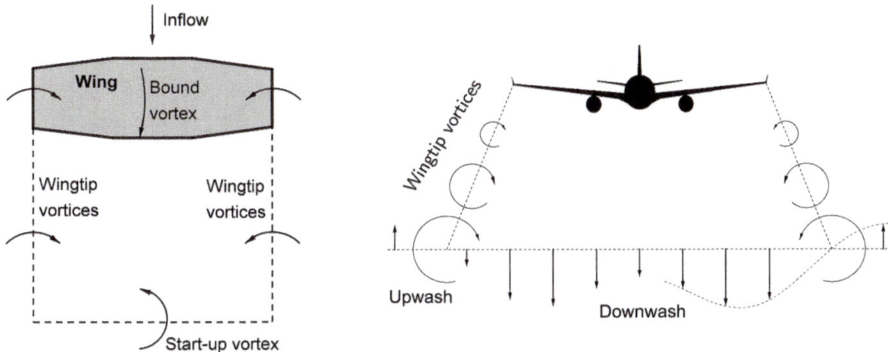

Fig. 6.3 (left) Top view of the generated vortex field. (right) Rear view.

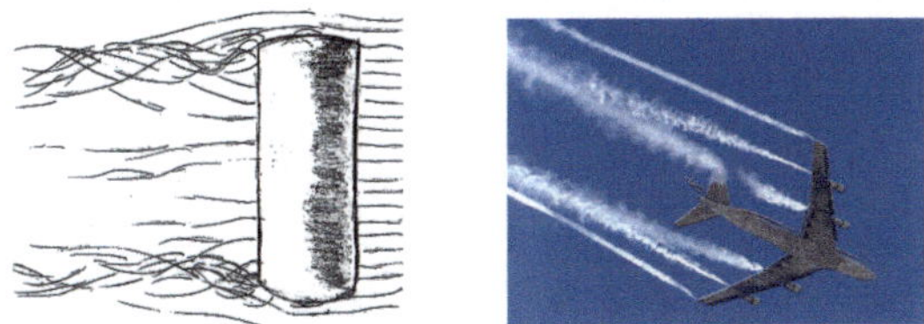

Fig. 6.4 (left) Cylinder in flow experiment. (right) Generation of vorticity on engines and wingtips of a Boeing 747. As this can destabilize following aircraft, a safety distance must be maintained at airports. Its disregard and pilot error led to the crash of an Airbus A300 in New York on November 12, 2001. Image source: NASA, Wikimedia Commons.

> Disregarding the initial vortex (start-up vortex), the subsequent vortex field is called a *horseshoe vortex*.

By embedding in the coordinate system of Fig. 6.1, the directions of vector quantities such as lift and drag are fixed and we can calculate them using their magnitudes. (As explained in Sect. 1.1, the magnitudes of vectors are set in normal print, for example $u = \|\mathbf{u}\|$.)

A force $\mathbf{F}$ acting on the wing is composed of its components $\mathbf{F}'(y)$ to the respective cross sections (Fig. 6.5). If we consider $\mathbf{F}'(y)$ as the *force density* on the chord line $-b/2 \leq y \leq b/2$, we obtain by integration

$$\mathbf{F} = \int_{-b/2}^{b/2} \mathbf{F}'(y)\,\mathrm{d}y$$

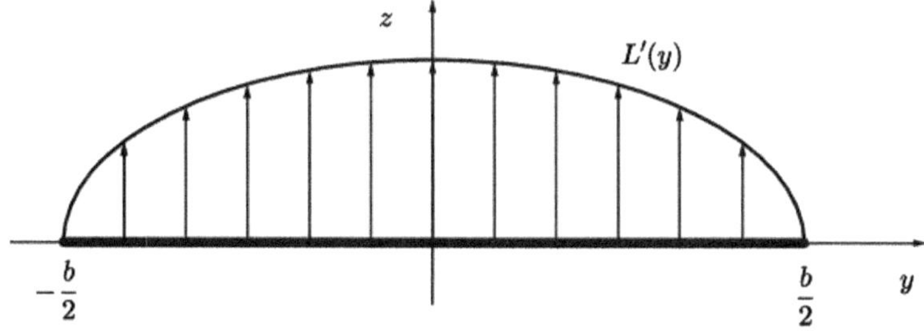

Fig. 6.5 Lift density on the chord line.

If the force density, as with lift and drag, always points in the same direction, then we can limit ourselves to the magnitudes in the integration and obtain

$$F = \int_{-b/2}^{b/2} F'(y)\,\mathrm{d}y \tag{6.1}$$

The *air density* in the undisturbed flow area is denoted by ϱ_∞.

Furthermore, there are
- u_∞ —*Freestream velocity* (inflow velocity)
- q_∞ —*Dynamic pressure*:

$$q_\infty = \frac{1}{2}\varrho_\infty u_\infty^2 \tag{6.2}$$

Profile calculations serve as the starting point for the three-dimensional theory. Figure 6.6 shows the profile at y in relation to force densities, wind speeds and angles.

Velocities in the profile plane (Fig. 6.6)

- $w(y)$ —*Downwash*
  Downward air speed above and below the wing, caused by pressure equalisation.
- $u_{\mathrm{res}}(y)$ —*Local relative wind speed*.
  It is the resultant of freestream velocity u_∞ and downwash.

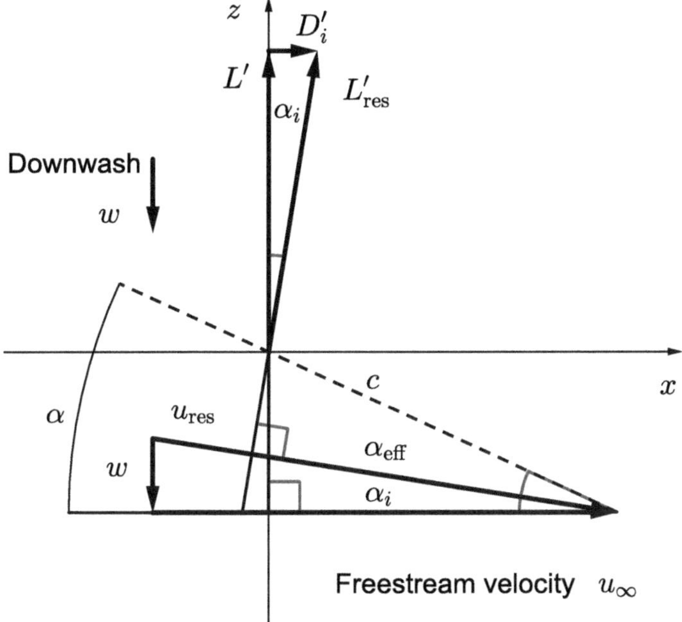

Fig. 6.6 Airfoil profile with wind speeds and force densities.

Angles in the profile plane (Fig. 6.6)

- $\alpha(y)$ —*Angle of attack*: Angle between chord line $c(y)$ and inflow speed u_∞

- $\alpha_{L=0}(y)$ —*Maximum angle of attack*. Here, the flow separates early and the lift decreases drastically. The value is determined experimentally and is tabulated for standardised profiles, see Abbott [1] and Example 6.3.

- $\alpha_i(y)$ —*Induced angle of attack*:
  Owing to the direction of w, it holds $\tan \alpha_i(y) = -w(y)/u_\infty$.
  Since α_i is small, it follows that $\alpha_i(y) \approx \tan \alpha_i(y)$, i.e.

$$\alpha_i(y) = -\frac{w(y)}{u_\infty} \tag{6.3}$$

- α_{eff} —*Effective angle of attack*.

$$\alpha_{\text{eff}} = \alpha - \alpha_i \tag{6.4}$$

Force densities in the profile plane (Fig. 6.6)

- $L'(y)$ —*Lift density.*
- $D'_i(y)$ —*Induced drag density.*

$$D'_i(y) = L'(y) \sin \alpha_i \qquad (6.5)$$

Forces on the (three-dimensional) airfoil   According to (6.1), we have

$$L = \int_{-b/2}^{b/2} L'(y)\, dy \qquad \textit{Lift} \qquad (6.6)$$

$$D_i = \int_{-b/2}^{b/2} D'_i(y)\, dy \qquad \textit{Induced drag} \qquad (6.7)$$

In addition, there is the *parasitic drag* D_0, which also occurs without lift in the flow and contains in particular the portion of the frictional forces.

$$D_0 = \frac{1}{2}\varrho S C_d u_\infty^2 \qquad (6.8)$$

This approximation formula was developed by Rayleigh and applies to air speeds between 25 and 330 m/s or Reynolds numbers Re $>$ 1000. The constant C_d is referred to as the *parasitic drag coefficient*. It depends not only on the shape of the obstacle, but also on the type of flow (laminar/turbulent), the viscosity and other parameters.

For low-speed aircraft, the *total drag* D is obtained as the sum

$$D = D_0 + D_i \qquad (6.9)$$

The force D_i is caused by vorticity generation, but unlike D_0 it does not contain any frictional forces.

As Fig. 6.6 illustrates, the downwash w induces a change in direction of the air flow in the two-dimensional vicinity of the airfoil profile. Instead of the freestream velocity u_∞, the flow behaviour near the airfoil is determined by the resultant u_{res}, which is reduced by α_i. Since the lift density is perpendicular to the flow velocity according to the Kutta-Joukowski theorem (Theorem 4.3), it experiences the same inclination. Consequently, we obtain the resulting force

$$L'_{\text{res}}(y) = \varrho u_{\text{res}}(y)\,\Gamma(y) \tag{6.10}$$

The inclination of the lift by α_i leads to an additional drag D_i, the so-called *induced drag*, which was missing in Sect. 4.2.

Problem 6.1 Prove that the lift formula

$$L'(y) = \varrho u_\infty \Gamma(y) \tag{6.11}$$

remains valid.

Problem 6.2 Derive the following formula for the induced drag.

$$D'_i(y) = \varrho w(y)\,\Gamma(y) \tag{6.12}$$

For the forces on the airfoil, it follows from (6.6), (6.7) that

$$L = \varrho_\infty u_\infty \int_{-b/2}^{b/2} \Gamma(y)\,dy \qquad \text{(Lift)} \tag{6.13}$$

$$D_i = \int_{-b/2}^{b/2} L'(y) \sin\alpha_i\,dy \qquad \text{(Induced drag)} \tag{6.14}$$

For similar airfoil geometry, both lift and drag can be converted proportionally to the wing dimensions. For this purpose, we introduce dimensionless quantities that can be determined by measurements on scaled-down models in the wind tunnel as a function of the angle of attack. For a given profile, the following value can be assigned to each cross section.

Sectional lift coefficient

$$C_l(y) = \frac{L'(y)}{q_\infty\,c(y)} \tag{6.15}$$

$$= \frac{2L'(y)}{\varrho_\infty u_\infty^2\,c(y)} = \frac{2\Gamma(y)}{u_\infty c(y)}$$

In the last step we used (6.11). By integrating along the lifting line, we determine the three-dimensional coefficients.

Fig. 6.7 NACA 64-210.

- *Lift coefficient*

$$C_L = \frac{L}{q_\infty S} = \frac{2L}{\varrho_\infty u_\infty^2 S} \qquad (6.16)$$

- *Induced drag coefficient*

$$C_{D_i} = \frac{D_i}{q_\infty S} = \frac{2D_i}{\varrho_\infty u_\infty^2 S} \qquad (6.17)$$

- *Total drag coefficient*

$$C_D = \frac{D}{q_\infty S} = \frac{2D}{\varrho_\infty u_\infty^2 S} \qquad (6.18)$$

Problem 6.3 Derive the formula for the total drag coefficient.

$$C_D = C_d + C_{D_i} \qquad (6.19)$$

Example 6.1 We determine the maximum angle of attack $\alpha_{L=0}$ for the profile NACA 64-210 (see Fig. 6.7). A corresponding diagram for $C_l = C_l(\alpha)$ can be found in Abbott [1], p. 564. From Sect. A.12, Fig. A.6 we obtain $\alpha_{L=0} = -1.8°$.

Calculation of Lift and Drag

In Sect. 4.4, we examined the dependence of lift on the angle of attack α for the Joukowski airfoil. It can be measured in the wind tunnel for any profile and is represented as curve $C_l = C_l(\alpha)$ (Fig. 6.8). According to the Joukowski model (4.64), we expect $C_l \sim \sin\alpha$. However, Fig. 6.8 shows a significant loss of lift already at $\alpha > 15°$, which is caused by vorticity generation on the upper side of the nose and subsequent flow separation. The involvement of vorticity complicates modelling with the complex potentials used. Therefore, the inaccuracy of (4.64) at increasing angles of attack is understandable.

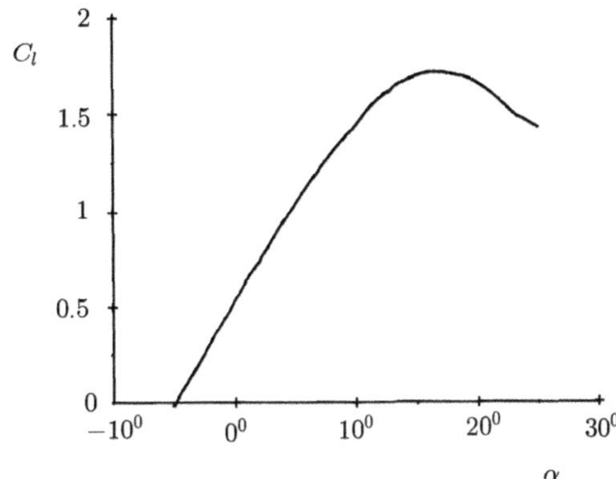

Fig. 6.8 Typical dependence of the coefficient C_l on the angle of attack α for measurements in the wind tunnel.

For permissible angles of attack, it holds

$$C_l(y) = a_0(\alpha_{\text{eff}} - \alpha_{L=0}) = a_0(\alpha - \alpha_i - \alpha_{L=0}) \tag{6.20}$$

with an experimentally determined constant $a_0 \approx 2\pi$. First, we use the approximation

$$C_l(y) = 2\pi(\alpha - \alpha_i - \alpha_{L=0}) \tag{6.21}$$

For small α we have $\sin\alpha \approx \alpha$. For this reason, Joukowski's formula (4.64) agrees well with the available experimental data. Tabulated values for C_l can be found on the page http://airfoiltools.com/airfoil/

Problem 6.4 Derive the following formulas.

$$C_L(y) = \frac{2}{u_\infty S} \int_{-b/2}^{b/2} \Gamma(y)\,dy \tag{6.22}$$

$$D_i \approx \int_{-b/2}^{b/2} L'(y)\alpha_i\,dy \approx \varrho_\infty u_\infty \int_{-b/2}^{b/2} \Gamma(y)\alpha_i\,dy \tag{6.23}$$

$$C_{D_i} = \frac{D_i}{q_\infty S} = \frac{2}{u_\infty S} \int_{-b/2}^{b/2} \Gamma(y)\alpha_i\,dy \tag{6.24}$$

Vorticity and Downwash

By inducing a downwash, the vorticity changes the flow direction. This relation will prove to be essential for the calculation of circulation.

Owing to the pressure balance at the wing tips, we have $L'(\pm b/2) = 0$. Because of (6.11), it follows that $\Gamma(\pm b/2) = 0$ and the circulation $\Gamma(y)$ decreases towards the wing tips. At y_0, it decreases by $\frac{d\Gamma}{dy}(y_0)$. According to Helmholtz's theorem (Theorem 5.4), this difference cannot disappear without a trace. In Prandtl's model, it is carried downstream by a trailing vortex, which starts at y_0 on the lifting line. Together with the bound vortex in the lifting line, the trailing vortices form the horseshoe vortex.

Using Biot-Savart's law (5.87), we obtain the downwash induced at point y_0 by the vortex filament through y.

$$dw(y_0) = -\frac{\frac{d\Gamma}{dy}\,dy}{4\pi(y_0 - y)}$$

Therefore, the downwash induced by the entire vortex field at y_0 is given by

$$w(y_0) = -\frac{1}{4\pi}\int_{-b/2}^{b/2} \frac{\frac{d\Gamma}{dy}\,dy}{y_0 - y} \tag{6.25}$$

According to the Kutta-Joukowski formula (4.13), the force acts perpendicular to its generating flow velocity. In particular, the lift density $L'_{\text{res}}(y)$ is perpendicular to $u_{\text{res}}(y)$ and the induced drag density $D'_i(y)$ is perpendicular to the downwash $w(y)$, see Fig. 6.6.

Because $w(y)$ is oriented[1] downwards, $D'_i(y)$ must be directed against the movement. It is a braking force. In contrast to the two-dimensional theory, where the vorticity does not induce a downwash $w(y)$, in the three-dimensional model there is drag even in the frictionless case.

Finally, it should be noted that the relative wind speed u_{res} is generally not horizontal and consequently the lift L does not have to be directed vertically upwards. Only the vertical component of L contributes to keeping the aircraft in the air.

Calculation for Elliptical Circulation

We examine the lift of a profile, whose circulation has an elliptical distribution (Fig. 6.9):

$$\Gamma(y) = \Gamma_0\sqrt{1 - \left(\frac{2y}{b}\right)^2} \tag{6.26}$$

---

[1] We will prove this direction in the special case of an elliptical circulation.

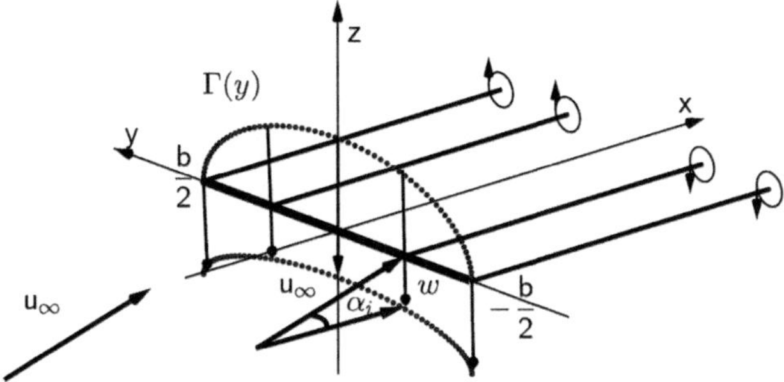

Fig. 6.9 Horseshoe vortex with lifting line $-\frac{b}{2} \leq y \leq \frac{b}{2}$ and trailing vortices.

Our goal is to find the wing geometry as well as the lift and drag. We use the substitution

$$y = \frac{b}{2}\cos\theta, \quad dy = -\frac{b}{2}\sin\theta \, d\theta, \tag{6.27}$$

where $y_1 = -\frac{b}{2}$ is transformed into $\theta_1 = \pi$ and $y_2 = \frac{b}{2}$ into $\theta_2 = 0$. From (6.26) it follows that

$$\Gamma(\theta) = \Gamma_0\sin\theta, \qquad 0 \leq \theta \leq \pi \tag{6.28}$$

Problem 6.5 Prove the formulas for an elliptical circulation distribution (6.26).

$$L = \varrho_\infty u_\infty \Gamma_0 \frac{b}{4}\pi \qquad \text{(Lift)} \tag{6.29}$$

$$C_L = \frac{b\Gamma_0\pi}{2u_\infty S} \qquad \text{(Lift coefficient)} \tag{6.30}$$

From (6.26) we get

$$\frac{d\Gamma}{dy} = -\frac{4\Gamma_0}{b^2}\frac{y}{\sqrt{1 - \frac{4y^2}{b^2}}}$$

By (6.25) it follows that

$$w(y_0) = \frac{\Gamma_0}{\pi b^2}\int_{-b/2}^{b/2}\frac{y}{\sqrt{1 - \frac{4y^2}{b^2}}}\frac{dy}{y_0 - y}$$

$$w(\theta_0) = -\frac{\Gamma_0}{2\pi b}\int_\pi^0 \frac{\cos\theta}{\cos\theta_0 - \cos\theta}\,d\theta = -\frac{\Gamma_0}{2\pi b}\int_0^\pi \frac{\cos\theta}{\cos\theta - \cos\theta_0}\,d\theta$$

This is the Glauert integral for the special case $n = 1$. Using Sect. A.7, formula (A.31), we obtain

$$\int_0^\pi \frac{\cos\theta}{\cos\theta - \cos\theta_0}\, d\theta = \frac{\pi \sin\theta_0}{\sin\theta_0} = \pi$$

Thus, we have the downwash

$$w(\theta_0) = -\frac{\Gamma_0}{2b} \quad \text{for } 0 \le \theta_0 \le \pi, \tag{6.31}$$

which (in this case) remains constant along the chord line. The sign indicates that w is directed downwards.

Now we determine the induced angle of attack and the drag. From (6.3) it follows that

$$\alpha_i(y) = -\frac{w(y)}{u_\infty} = \frac{\Gamma_0}{2bu_\infty} \tag{6.32}$$

The value remains constant along the chord line. Using (6.24) and (6.26), we find the coefficient of induced drag.

$$\begin{aligned}
C_{D_i} &= \frac{2}{u_\infty S}\int_{-b/2}^{b/2} \Gamma(y)\alpha_i\, dy = \frac{2\alpha_i\Gamma_0}{u_\infty S}\int_{-b/2}^{b/2}\sqrt{1 - \left(\frac{2y}{b}\right)^2}\, dy \\
&= \frac{\alpha_i\Gamma_0 b}{u_\infty S}\int_0^\pi \sin^2\theta\, d\theta \quad \text{by (6.27)} \\
&= \frac{\alpha_i b\Gamma_0\pi}{2u_\infty S}
\end{aligned}$$

Taking into account (6.30), (6.32) and the aspect ratio $\mathcal{R} = b^2/S$, we obtain

$$C_{D_i} = \frac{C_L^2}{\pi\,\mathcal{R}} \tag{6.33}$$

In (6.53) we show the relation $C_{D_i} \sim C_L^2/\mathcal{R}$ for arbitrary wings, which allows the following conclusions.

Fig. 6.10 Shy albatross (Thalassarche cauta). With a mass of 4.1 kg, the wingspans range between 220 and 256 cm. The birds can travel up to 1000 km per day.

- While maintaining the wing proportions, we have $C_{D_i} \sim C_L^2$, i.e. the energy expenditure increases disproportionately with increased lift.
- Owing to $C_{D_i} \sim 1/\!R$, wing designs with large $\!R$, i.e. long and narrow wings, prove to be cost-effective.

Example 6.2 With a weight of up to 12 kg and a wingspan of over 3.5 m, albatrosses are among the largest flying birds. We will show in Sect. 7.1 that the weight requires a high energy expenditure, which almost limits their flight technique to gliding. Elongated, narrow wings help to minimize the induced drag (Fig. 6.10).

Using the above formulas, we can determine the wing geometry for elliptical circulation distributions. Because of (6.32), the induced angle of attack α_i is independent of y. From (6.21) it follows that

$$C_l(y) = 2\pi(\alpha - \alpha_i - \alpha_{L=0}) \quad \text{constant for all } y. \tag{6.34}$$

From (6.15) and (6.26) we find that

$$C_l = \frac{2\Gamma_0}{u_\infty c(y)} \sqrt{1 - \left(\frac{2y}{b}\right)^2} \tag{6.35}$$

and obtain the ellipse equation of the camber line $c(y)$

$$\left(\frac{C_l u_\infty}{2\Gamma_0} c(y)\right)^2 + \left(\frac{2}{b} y\right)^2 = 1$$

Since the flow conditions are uniquely determined for a given inflow and wing geometry, wings with an elliptical planform must have an elliptical circulation distribution (Fig. 6.11).

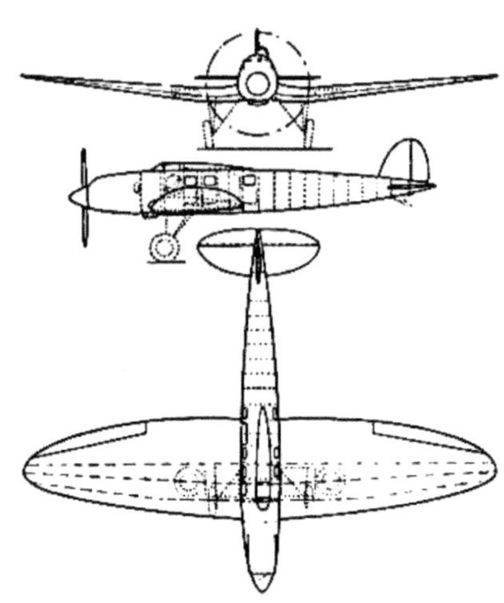

Fig. 6.11 The Heinkel He 70 Blitz was a German mail and passenger aircraft, which was built from 1932. It could carry five passengers in addition to the pilot and was at times the fastest commercial aircraft in the world. The elliptical wings used for the first time contributed to its exceptional aerodynamic properties.

6.2   Elliptical Wings

We consider a wing with an elliptical planform and the camber line

$$c(y) = c_r \sqrt{1 - \left(\frac{2y}{b}\right)^2} \tag{6.36}$$

Owing to the remark at the end of the previous section, we can use elliptical circulation (6.26) and all resulting conclusions.

First, we calculate Γ_0 from the wing geometry. In (6.35) we replace (6.34) for C_l, (6.32) for α_i and (6.36) for $c(y)$. After simplification we get

$$\pi\left(\alpha - \frac{\Gamma_0}{2bu_\infty} - \alpha_{L=0}\right) = \frac{\Gamma_0}{u_\infty c_r}$$

$$\Gamma_0 = \frac{2\pi u_\infty(\alpha - \alpha_{L=0})}{\frac{2}{c_r} + \frac{\pi}{b}} \tag{6.37}$$

Using (6.30) and (6.33) we now find the specific coefficients C_L and C_{D_i}. The following example is taken from [4].

Example 6.3 We examine a wing with the following parameters

- profile NACA 64-210
- elliptical planform, wingspan $b = 10\,\text{m}$, wing root $c_r = 2.5\,\text{m}$
- wing mass $m_f = 1000\,\text{kg}$
- freestream velocity $u_\infty = 50\,\text{m/s}$, inflow angle $\alpha = 8°$

The following quantities are to be calculated:

- lift L, induced drag D_i
- max. payload m_n
- max. acceleration a (with no payload) at sea level
  (air density $\varrho_\infty = 1.225\,\text{kg/m}^3$)

The maximum angle of attack $\alpha_{L=0} = -1.8°$ was read off in Example 6.1. From (6.37) it follows

$$\Gamma_0 = \frac{2\pi \cdot 50\,\text{m/s} \cdot (8° - (-1.8°)) \cdot \frac{\pi}{180°}}{\frac{2}{2.5\,m} + \frac{\pi}{10\,m}} = 48.23\,\text{m}^2/\text{s}$$

The ellipse area results from the semi-axes $b/2$ and $c_r/2$.

$$S = \pi \frac{b}{2} \frac{c_r}{2} \tag{6.38}$$

$$= 19.63\,\text{m}^2$$

For the aspect ratio we find

$$\mathit{R} = \frac{b^2}{S} = 5.09$$

By (6.30) and (6.33) we obtain

the lift coefficient
$$C_L = \frac{10\,\text{m} \cdot 48.23\,\text{m}^2/\text{s} \cdot \pi}{2 \cdot 50\,\text{m/s} \cdot 19.63\,\text{m}^2} = 0.77$$

the induced drag coefficient
$$C_{D_i} = \frac{0.77^2}{\pi \cdot 5.09} = 0.037$$

Using (6.16), we calculate the lift at sea level

$$L = \frac{1}{2}\varrho_\infty u_\infty^2 S C_L = \frac{1}{2} \cdot 1.225\,\text{kg/m}^3 \cdot (50\,\text{m/s})^2 \cdot 19.63\,\text{m}^2 \cdot 0.77 = 23.1\,\text{kN}$$

and by (6.17) the induced drag

$$D_i = \frac{1}{2}\varrho_\infty u_\infty^2 S C_{D_i} = \frac{1}{2} \cdot 1.225\,\text{kg/m}^3 \cdot (50\,\text{m/s})^2 \cdot 19.63\,\text{m}^2 \cdot 0.037 = 1.1\,\text{kN}$$

The lifting force F_h is directed vertically upwards and can be determined from the right-angled force triangle $L^2 = F_h^2 + D_i^2$.

$$F_h = \sqrt{L^2 - D_i^2} = \sqrt{23.1^2 - 1.1^2}\,\text{kN} = 23.07\,\text{kN}$$

The maximum payload m_n results from

$$F_h = (m_n + m_f)g$$

$$m_n = \frac{F_h}{g} - m_f = \frac{23.07\,\text{kN}}{9,81\,\text{m/s}^2} - 1000\,\text{kg} = 1350\,\text{kg}$$

For the maximum acceleration a, it follows from $m_f\,a = F_h - m_f\,g$ that

$$a = \frac{F_h}{m_f} - g = 23.07\,\text{m/s}^2 - 9.81\,\text{m/s}^2 = 13.26\,\text{m/s}^2.$$

It exceeds the acceleration due to gravity by a factor of 1.35 ($= \frac{13.26}{9.81}$).

Figures 6.12 and 6.13 show the dependence of the coefficients C_L and C_{D_i} on the angle of attack $-10° \leq \alpha \leq 10°$.

If we examine the specific coefficients in relation to the aspect ratio $\mathcal{R} = b^2/S$, we can find an optimal design. First, we replace all lengths by expressions with b or S. From (6.37) and (6.38) it follows that

$$\Gamma_0 = \frac{2\pi u_\infty (\alpha - \alpha_{L=0})}{\frac{b\pi}{2S} + \frac{\pi}{b}}$$

Fig. 6.12 Lift coefficient C_L for an elliptical wing ($b = 10\,\text{m}$, $c_r = 2.5\,\text{m}$, $u_\infty = 50\,\text{m/s}$).

Fig. 6.13 Drag coefficient C_{D_i} for an elliptical wing ($b = 10\,\text{m}$, $c_r = 2.5\,\text{m}$, $u_\infty = 50\,\text{m/s}$).

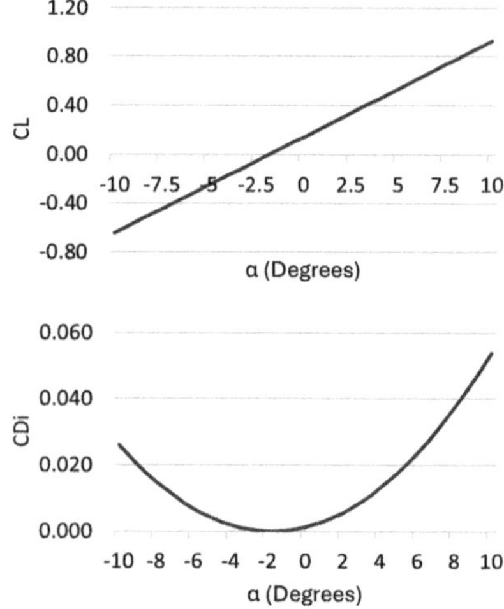

Using (6.30), we get the lift coefficient

$$
C_L = \frac{b\pi}{2u_\infty S}\,\frac{2\pi u_\infty(\alpha - \alpha_{L=0})}{\frac{b\pi}{2S} + \frac{\pi}{b}} = \frac{b\pi}{S}\,\frac{\alpha - \alpha_{L=0}}{\frac{b}{2S} + \frac{1}{b}} = \frac{2\pi(\alpha - \alpha_{L=0})}{1 + \frac{2S}{b^2}}
$$

$$
= \frac{2\pi(\alpha - \alpha_{L=0})}{1 + \frac{2}{\!R}} \tag{6.39}
$$

For the induced drag coefficient, we obtain from (6.33) and (6.39)

$$
C_{D_i} = \frac{1}{\pi\!R}\left(\frac{2\pi(\alpha - \alpha_{L=0})}{1 + \frac{2}{\!R}}\right)^2
$$

$$
= \frac{4\pi(\alpha - \alpha_{L=0})^2}{\!R + 4 + \frac{4}{\!R}} \tag{6.40}
$$

With increasing $\!R$, i.e. longer and narrower wings of the same area, the lift coefficient C_L increases and and reaches the corresponding value of the two-dimensional theory in the limiting case $\!R \to \infty$. The coefficient C_{D_i} has an extremum depending on $\!R$. Since

$$
\frac{\mathrm{d}C_{D_i}}{\mathrm{d}\!R} = \frac{-4\pi(\alpha - \alpha_{L=0})^2}{(\!R + 4 + \frac{4}{\!R})^2}\left(1 - \frac{4}{\!R^2}\right),
$$

the necessary condition $1 - \frac{4}{\!R^2} = 0$ results in a maximum at $\!R = 2$, i.e. for $b = \frac{c_r}{2}\pi$. From $\!R > 2$ the induced drag decreases continuously and disappears for $\!R \to \infty$ as in the two-dimensional case. The following explains the pronounced contrails on the wingtips in Fig. 6.4.

Problem 6.6 Using elliptical wings as an example, show that the strength of the trailing vortices is maximum at the wing tips.

Finally, we would like to point out a result that Prandtl [5] was able to derive for an elliptical circulation. It allows an interpretation of lift via momentum interactions.
   After eliminating Γ_0, it follows from (6.29) and (6.31) that

$$
\frac{1}{2}\varrho_\infty u_\infty \pi b^2\, w = -L \tag{6.41}
$$

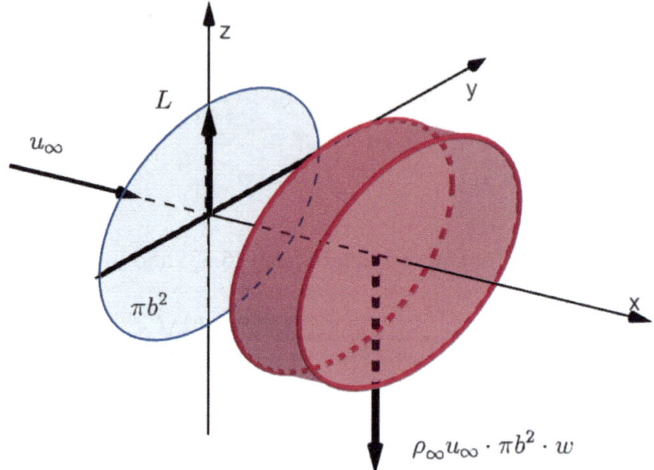

Fig. 6.14 The air flowing through the circular surface per unit of time is given a downward impulse by the downwash.

We interpret πb^2 as a circular area with the wingspan as the diameter. Then

$$M = \varrho_\infty u_\infty \pi b^2$$

is the mass of air flowing through this circular area per unit time (Fig. 6.14). The quantity $Mw = \varrho_\infty u_\infty \pi b^2 w$ corresponds to the momentum that the flowing mass receives due to the downwash. Half of this momentum becomes effective as lift for the wing.

6.3 Prandtl's Fundamental Equation

For a given wing, we first look for the circulation. All other parameters can then be found using the known formulas.

Derivation of the Fundamental Equation

The circulation $\Gamma(y)$ depends on the given values for the angle of attack α and the inflow speed u_∞. In addition, we need the experimentally determined values for the maximum angle of attack $\alpha_{L=0}$ and the constant $a_0 = dC_l/d\alpha$ in (6.20).

Theorem 6.1 *The circulation $\Gamma(y)$ satisfies the integro-differential equation*

$$\alpha(y_0) = \frac{2\Gamma(y_0)}{a_0 u_\infty c(y_0)} + \alpha_{L=0}(y_0) + \frac{1}{4\pi u_\infty} \int_{-b/2}^{b/2} \frac{\frac{d\Gamma}{dy} dy}{y_0 - y}, \quad -\frac{b}{2} \le y_0 \le \frac{b}{2}$$

$$(6.42)$$

Proof From (6.15) and (6.20) we find that

$$\frac{\Gamma(y)}{u_\infty c(y)} = \frac{a_0}{2}(\alpha - \alpha_i - \alpha_{L=0})$$

By (6.25), (6.3) we obtain the induced angle of attack

$$\alpha_i(y_0) = \frac{1}{4\pi u_\infty} \int_{-b/2}^{b/2} \frac{\frac{d\Gamma}{dy}\,dy}{y_0 - y}, \tag{6.43}$$

which is substituted into the above equation. Hence

$$\frac{\Gamma(y)}{u_\infty c(y)} = \frac{a_0}{2}\left(\alpha - \frac{1}{4\pi u_\infty} \int_{-b/2}^{b/2} \frac{\frac{d\Gamma}{dy}\,dy}{y_0 - y} - \alpha_{L=0}\right),$$

which gives (6.42). $\square$

Optimal Shape of an Elliptical Wing

We want to show that losses due to induced drag are minimised in elliptical wing design. As in the elliptical special case, we parameterize the lifting line of the wing by (6.27)

$$y = \frac{b}{2}\cos\theta, \ \ dy = -\frac{b}{2}\sin\theta\,d\theta \quad \text{with} \quad y_1 = -\frac{b}{2} \text{ at } \theta_1 = \pi, \quad y_2 = \frac{b}{2} \text{ at } \theta_2 = 0$$

For evaluation we need the well-known formula

Problem 6.7 Prove

$$\int_0^\pi \sin(nx)\,\sin(mx)\,dx = \begin{cases} \pi/2 & \text{if } m = n \\ 0 & \text{if } m \neq n \end{cases} \tag{6.44}$$

When solving the fundamental equation, we use the finite Fourier series approach, in which only the components of the sine functions occur.

$$\Gamma(\theta) = 2bu_\infty \sum_{n=1}^{N} A_n \sin(n\theta) \tag{6.45}$$

The restriction to sine components is necessary to ensure the disappearance of circulation and lift at the wing tips.

$$\Gamma(\theta) = 0 \qquad \text{for } \theta = 0 \text{ or } \theta = \pi$$

In (6.45) the elliptical wing (6.28) proves to be the special case $N = 1$. Based on this, all other terms $n \geq 2$ can be interpreted as a perturbation. By (6.31) we had shown the direction of the downwash w for elliptical wings. Small disturbances will maintain the direction.

Let us evaluate the integral term of (6.42).

$$\frac{d\Gamma}{dy} = \frac{d\Gamma}{d\theta}\frac{d\theta}{dy} = 2bu_\infty \sum_{n=1}^{N} n A_n \cos(n\theta)\frac{d\theta}{dy}$$

$$\int_{-b/2}^{b/2} \frac{\frac{d\Gamma}{dy}dy}{y_0 - y} = 2bu_\infty \int_{\pi}^{0} \frac{\sum_{n=1}^{N} n A_n \cos(n\theta)\,d\theta}{\frac{b}{2}(\cos\theta_0 - \cos\theta)}$$

$$= 4u_\infty \sum_{n=1}^{N} n A_n \int_{0}^{\pi} \frac{\cos(n\theta)}{\cos\theta - \cos\theta_0}\,d\theta$$

Using formula (A.31), we obtain for the Glauert integral

$$\int_{0}^{\pi} \frac{\cos(n\theta)}{\cos\theta - \cos\theta}\,d\theta = \frac{\pi\sin(n\theta)}{\sin\theta}$$

and thus

$$\int_{-b/2}^{b/2} \frac{\frac{d\Gamma}{dy}dy}{y_0 - y} = 4u_\infty \sum_{n=1}^{N} n A_n \frac{\pi\sin(n\theta)}{\sin\theta} = \frac{4\pi u_\infty}{\sin\theta}\sum_{n=1}^{N} n A_n \sin(n\theta) \qquad (6.46)$$

If we insert the latter together with (6.45) into (6.42), we get

$$\alpha(\theta_0) = \frac{4bu_\infty \sum_{n=1}^{N} A_n \sin(n\theta_0)}{a_0 u_\infty c(\theta_0)} + \alpha_{L=0}(\theta_0) + \frac{1}{4\pi u_\infty}\frac{4\pi u_\infty}{\sin\theta_0}\sum_{n=1}^{N} n A_n \sin(n\theta_0)$$

We drop the index at θ_0 and obtain after rearrangement

$$\alpha(\theta) = \frac{4b}{a_0 c(\theta)}\sum_{n=1}^{N} A_n \sin(n\theta) + \alpha_{L=0}(\theta) + \frac{1}{\sin\theta}\sum_{n=1}^{N} n A_n \sin(n\theta)$$

We introduce the parameter

$$\mu = \frac{a_0 c(\theta)}{4b} \qquad (6.47)$$

and obtain

$$\alpha(\theta) - \alpha_{L=0}(\theta) = \frac{1}{\mu} \sum_{n=1}^{N} A_n \sin(n\theta) + \frac{1}{\sin\theta} \sum_{n=1}^{N} n A_n \sin(n\theta)$$

$$\mu \sin\theta \left[\alpha(\theta) - \alpha_{L=0}(\theta)\right] = \sum_{n=1}^{N} A_n \sin(n\theta) \left[\sin\theta + \mu n\right] \tag{6.48}$$

By specifying the values $\theta_1, \ldots, \theta_N$, we obtain a linear system of equations from (6.48) for the desired coefficients $A_1, \ldots, A_N$.

Using (6.22), we find the lift coefficient. After inserting (6.45) and (6.27), we get

$$C_L(\theta) = \frac{2b^2}{S} \sum_{n=1}^{N} A_n \int_0^{\pi} \sin(n\theta) \, \sin\theta \, d\theta$$

Owing to (6.44), only the term for $n = 1$ contributes. Hence

$$C_L = \frac{b^2 \pi A_1}{S} = A_1 \pi \, \mathit{R} \tag{6.49}$$

Now we determine the induced angle of attack and the induced drag coefficient. If we substitute (6.46) into (6.43), we get

$$\alpha_i(y_0) = \frac{1}{\sin\theta} \sum_{n=1}^{N} n A_n \sin(n\theta) \tag{6.50}$$

After substituting (6.50), (6.45) and (6.27) into (6.24), we obtain the induced drag coefficient.

$$C_{D_i} = \frac{2}{S} \int_{\pi}^{0} \left[b \sum_{n=1}^{N} A_n \sin(n\theta) \right] \left[\sum_{n=1}^{N} n A_n \sin(n\theta) \, (-b) \right] d\theta$$

$$= \frac{2b^2}{S} \int_0^{\pi} \left[\sum_{n=1}^{N} A_n \sin(n\theta) \right] \left[\sum_{m=1}^{N} m A_m \sin(m\theta) \right] d\theta$$

$$= 2\mathit{R} \int_0^{\pi} \left[\sum_{n=1}^{N} \sum_{m=1}^{N} A_n A_m \sin(n\theta) \sin(m\theta) \right] d\theta$$

Owing to (6.44), only the terms for $m = n$ contribute and we find that

$$C_{D_i} = \mathit{\text{Æ}} \pi \sum_{n=1}^{N} n A_n^2$$

To examine C_{D_i} as a function of C_L, we introduce the following quantities.

$$\delta = \sum_{n=2}^{N} n \left(\frac{A_n}{A_1} \right)^2 \tag{6.51}$$

$$e = \frac{1}{1 + \delta} \qquad \text{Oswald efficiency number} \tag{6.52}$$

(Here we use italics for e in contrast to Euler number e.) It follows that

$$C_{D_i} = \pi \mathit{\text{Æ}} \left[A_1^2 + \sum_{n=2}^{N} n A_n^2 \right] = \pi \mathit{\text{Æ}} A_1^2 \left[1 + \sum_{n=2}^{N} n \left(\frac{A_n}{A_1} \right)^2 \right]$$

$$= \pi \mathit{\text{Æ}} A_1^2 (1 + \delta) = \frac{C_L^2}{\pi \mathit{\text{Æ}}} (1 + \delta) \quad \text{with (6.51), (6.49)}$$

$$= \frac{C_L^2}{\pi e \mathit{\text{Æ}}} \tag{6.53}$$

An energetically favourable design is achieved by reducing the specific induced drag C_{D_i}, i.e. for the largest possible e or the smallest possible δ. The optimum is found at $e = 1$ or $\delta = 0$. Owing to (6.51), we have $A_n = 0$ for $n \geq 2$ and the circulation (6.45) becomes elliptical. Consequently, the elliptical wing shape minimises the losses due to induced drag.

Wings with Symmetric Mass Distribution

We use an example to show how to calculate the circulation for a symmetric wing with a given camber line $c(y)$ and span b.

A wing is called *twisted* if the angle of attack depends on the position y on the chord line (see Fig. 6.15).

For an *untwisted wing* with a given inflow, both quantities α and $\alpha_{L=0}$ remain constant along the chord line.

Fig. 6.15 Twisted wing.

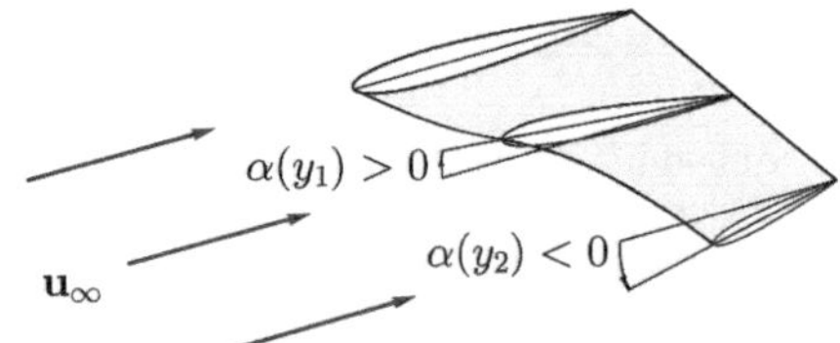

Problem 6.8 Prove that in (6.45) only functions $\sin(n\theta)$ with odd n occur if the circulation is symmetric with respect to $y = 0$.

Example 6.4 Using the approach (6.45) for $N = 5$, we model an untwisted wing with a camber line $c =$ constant and a symmetrical mass distribution.

First, we look for a linear system of equations for the coefficients A_i ($i = 1, \ldots, 5$) in (6.45). Under the given conditions, the circulation (6.45) is symmetric with respect to $y = 0$. Owing to Problem 6.8, we find that $A_2 = A_4 = 0$. To calculate the remaining coefficients A_1, A_3, A_5, we need three values for θ, which can be chosen as follows.

$$\theta_1 = \frac{\pi}{2}, \qquad \theta_2 = \frac{\pi}{3}, \qquad \theta_3 = \frac{\pi}{6}$$

From (6.48) we obtain the system

$$M\mathbf{A} = \mathbf{B}, \tag{6.54}$$

where the vector $\mathbf{A} = [A_1, A_3, A_5]$ contains the relevant unknowns for (6.45). The matrix

$$M = \begin{bmatrix} \sin\theta_1(\sin\theta_1 + \mu) & \sin(3\theta_1)(\sin\theta_1 + 3\mu) & \sin(5\theta_1)(\sin\theta_1 + 5\mu) \\ \sin\theta_2(\sin\theta_2 + \mu) & \sin(3\theta_2)(\sin\theta_1 + 3\mu) & \sin(5\theta_2)(\sin\theta_2 + 5\mu) \\ \sin\theta_3(\sin\theta_3 + \mu) & \sin(3\theta_3)(\sin\theta_1 + 3\mu) & \sin(5\theta_3)(\sin\theta_3 + 5\mu) \end{bmatrix}$$

and the vector

$$\mathbf{B} = \begin{bmatrix} \mu \sin\theta_1 (\alpha - \alpha_{L=0}) \\ \mu \sin\theta_2 (\alpha - \alpha_{L=0}) \\ \mu \sin\theta_3 (\alpha - \alpha_{L=0}) \end{bmatrix}$$

are given. (Since b, c and $a_0 = 2\pi$ are known, we calculate μ by (6.47), while $\alpha_{L=0}$ can be determined as in Example 6.1. The angle α takes on any desired value.)

References and Further Reading

1. Abbott I.H., von Doenhoff A.E.: *Theory of Wing Sections. Including a Summary of Airfoil Data.* Dover, New York (1959)
2. Anderson J.: *Fundamentals of Aerodynamics.* McGraw-Hill Education, New York (2017)
3. Childress S.: *An Introduction to Theoretical Fluid Mechanics.* Courant Lecture Notes in Mathematics Vol. 19, AMS, Providence RI (2009)
4. Gutmark E.J.: *Applied Aerodynamics - Finite Wing Theory* (Script) http://www.ase.uc.edu/class/ AEEM456/Section_2_Notes.pdf Accessed 3rd Sept 2024
5. Prandtl L., Betz A.: *Vier Abhandlungen zur Hydrodynamik und Aerodynamik.* Göttinger Klassiker der Strömungsmechanik, Band 3, Universitätsverlag Göttingen, Göttingen (2010)
6. Prandtl L.,Tietjens O.G.: *Fundamentals of Hydro- and Aeromechanics.* Dover Publications, New York (1957)

7.1 Energy Requirement for Gliding and Hovering

Birds and insects have developed different flight techniques. Heavy birds like albatrosses can glide with barely any wing movement. However, hovering is only possible for insects and hummingbirds. Only a few species, including kingfishers and buzzards, master the similar "wind hovering", usually with supportive headwind. Other techniques are used for special manoeuvres, e.g. during landing or steep take-off (see [7]). Using the example of gliding and hovering, we want to show that these techniques are only possible for certain masses, with hovering flight most strongly limiting the permissible mass.

Let us first consider *hovering*. When the wings beat, air is sucked in from above and pumped downwards. Figure 7.1 demonstrates the principle of ring vortex formation, which creates lift during downstream flow. When the wings beat, the air receives an impulse that increases its kinetic energy. We want to calculate this energy gain. The following model [5] reduces the complicated process of vortex formation and separation on moving wings to the flow through the area S in the xy-plane (Fig. 7.2), which corresponds to the wings in a horizontal position. We assume that the momentum transfer to the air occurs immediately after it passes through S.

Problem 7.1 Let ϱ be the air density and m_i the air mass passing through the area S per unit of time. Derive the formulas

$$S = \phi r^2 \tag{7.1}$$

$$\Delta I = \phi r^2 \varrho v_i v_w \qquad \text{momentum transferred to } m_i \tag{7.2}$$

$$P_i = \frac{1}{2}\varrho\phi r^2 v_i v_w^2 \qquad \text{kinetic energy of } m_i \text{ in the wake} \tag{7.3}$$

The original version of the chapter has been revised. A correction to this chapter can be found at https://doi.org/10.1007/978-3-662-72240-4_9

© The Author(s), under exclusive license to Springer-Verlag GmbH, DE, part of Springer Nature 2026, corrected publication 2026
R. Spielmann, *Theoretical Fluid Mechanics*,
https://doi.org/10.1007/978-3-662-72240-4_7

Fig. 7.1 When hovering, the wings perform a characteristic turn in addition to flapping. Each wing beat creates a horseshoe vortex, which is initially bound to the wings. After detachment, it closes and drifts downwards as a ring vortex. These ring vortices are responsible for generating lift. The downward airflow is called wake and has the speed v_w.

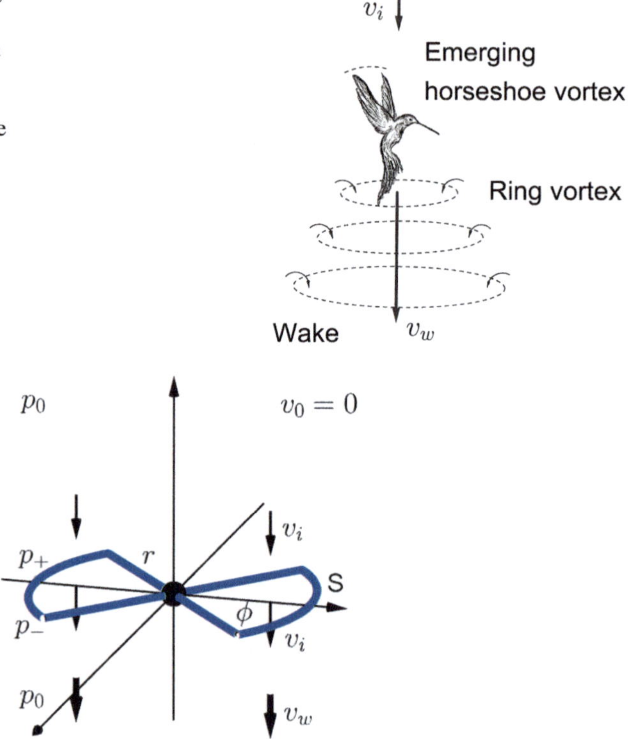

Fig. 7.2 Simplified flow model through a horizontal plane. We assume both wings to be circular sectors with radius r and angle ϕ. At a sufficient height above the wings, there is calm $v_0 = 0$ at atmospheric air pressure p_0. The air is pumped through the area S at speed v_i. The pressure p_+ and p_- prevails immediately above and below S. In the wake, the air flows at v_w, while the air pressure has normalised to p_0.

Let us determine the flow velocities. To the hovering state applies Newton's second law, i.e. the rate of change of momentum of the air (7.2) must be equal to the weight of the bird.

$$mg = \phi r^2 \varrho v_i v_w \tag{7.4}$$

When passing through the area S, the wing does work on the air, so that Bernoulli's theorem (Theorem 3.2) is not applicable. However, it can be used before or after the passage. We apply it to the flow above S. From (3.33), taking into account $v_0 = 0$, we get

$$p_0 - p_+ = \frac{1}{2}\varrho v_i^2 \tag{7.5}$$

Applying it to the flow below S, we obtain

$$p_- - p_0 = \frac{1}{2}\varrho(v_w^2 - v_i^2) \tag{7.6}$$

The addition of both equations yields

$$p_- - p_+ = \frac{1}{2}\varrho v_w^2 \tag{7.7}$$

Now, we find the force generated by the pressure on $S = \phi r^2$.

$$F = \left\| \int_S p\,\mathbf{n}\,dS \right\| = (p_- - p_+)\phi r^2 = \frac{1}{2}\varrho v_w^2 \phi r^2$$

When hovering, it balances the weight of the bird, i.e. we have

$$mg = \frac{1}{2}\varrho v_w^2 \phi r^2 \tag{7.8}$$

Problem 7.2 Derive the following formulas.

$$v_w = 2v_i \tag{7.9}$$

$$P_i = \sqrt{\frac{m^3 g^3}{2\varrho r^2 \phi}} \tag{7.10}$$

Since the surrounding air was set in motion exclusively by the bird, P_i is a measure of the animal's *energy consumption per unit of time while hovering*.

To estimate the energy balance, we examine the physical quantities of a body after similarity transformation with the scaling factor r. For lengths l, areas S and volumes V, we get the proportionalities

$$l \sim r, \qquad S \sim r^3, \qquad V \sim r^3$$

Since $m \sim V \sim r^3$, it follows from (7.10) that

$$P_i \sim \left(\frac{m^3}{r^2}\right)^{\frac{1}{2}} \sim r^{\frac{7}{2}} \qquad \text{energy consumption while hovering} \tag{7.11}$$

We are looking for the corresponding estimate for *gliding*. For simplicity, we assume an elliptical wing shape. From (6.37), we get for the circulation

$$\Gamma_0 \sim \frac{1}{\frac{1}{r} + \frac{1}{r}} \sim r$$

Fig. 7.3 Aeshna constricta. By using two pairs of counter-rotating wings at the same frequency, the dragonfly can always generate the crucial downward stroke for propulsion. The elongated abdomen serves as a stabilizer. Photo template by Erik Gauger www.notesfromtheroad.com.

Using (6.29), we obtain for the lift

$$L \sim \Gamma_0 r \sim r^2$$

Since L is proportional to the required power P, it follows that

$$P \sim r^2 \qquad \text{energy consumption while gliding} \qquad (7.12)$$

The comparison of (7.11) with (7.12) for increasing length $r \to \infty$ shows that the energy requirement for hovering increases faster than for gliding. Therefore, albatrosses, among which are the largest flying birds, are almost limited to gliding.[1] This technique has not prevailed in small birds and insects. Since their behavior in gliding flight is determined by smaller Reynolds numbers, the risk of early flow separation increases and endangers flight stability. (see Sect. 5.5).

For small birds and insects, hovering is energetically feasible. Dragonflies can even fly backwards (Fig. 7.3). When they suddenly enter free fall due to turbulence, they perform maneuvers similar to a backflip to return to a straight flight position (see [2]).

7.2 Torques

The outbreak of the First World War sparked a strong interest in aviation, which rapidly advanced its development. Novel applications revealed aerodynamic effects that gave unexpected impetus to young research. As a counterweight to allied biplanes, the German Fokker factories developed the monoplane Fokker D.VIII, which was characterised by a lower aerodynamic drag (Fig. 7.5). In accordance

---

[1] For animals, taking off by gliding is difficult and must therefore either be replaced by other techniques or supported by aids. Wandering albatrosses, which weigh up to 12 kg, require a wind speed of at least 12 km/h and a long run-up to take off. When landing in gliding flight at high speed they can roll over.

with the contemporary design, the wing structure was very light and consisted of two parallel wooden beams in the span direction, which were connected by cross beams and covered with fabric.

In combat, the inadequately tested fighter aircraft caused a shock when the wing suddenly broke. After a series of fatal accidents, the military authorities ordered an investigation. Among other things, it was found that the wings are subjected to considerable torques.[2] We want to calculate them using the simplified model of a flow around a plate.

The starting point is the circular potential (5.51), (5.52), which is transferred to the exterior of the plate using the Joukowski transformation (5.46), i.e.

$$w(\zeta) = u_\infty\left[\zeta e^{-i\alpha} + \frac{e^{i\alpha}}{\zeta}\right] - \frac{i\Gamma}{2\pi}\ln\zeta, \quad |\zeta| \geq 1 \tag{7.13}$$

$$J(\zeta) = \frac{1}{2}\left(\zeta + \frac{1}{\zeta}\right) \qquad \text{coordinate transformation} \tag{7.14}$$

To ensure the Joukowski condition, the circulation should be

$$\Gamma = -2\pi u_\infty \sin\alpha \tag{7.15}$$

As usual, we denote the plate potential by

$$\hat{w}(z) = w(J^{-1}(z))$$

The torque about the origin, which is also the centre of the plate, is determined according to the Blasius theorem (Theorem 4.5)

$$M = -\frac{\varrho}{2}\,\text{Re}\left[\int_C z\left(\frac{d\hat{w}}{dz}\right)^2 dz\right] \tag{7.16}$$

Problem 7.3 Prove the formula

$$z\left(\frac{d\hat{w}}{dz}\right)^2 dz = \left(\frac{dw}{d\zeta}\right)^2\left(\zeta + \frac{2}{\zeta} + \frac{2}{\zeta^3} + \frac{2}{\zeta^5} + \dots\right)d\zeta \qquad \text{for } |\zeta| > 1 \tag{7.17}$$

Theorem 7.1 *A horizontal plate is flowed at an angle α (Fig. 7.4). Then it receives a clockwise torque with respect to the centre*

$$M = -4\varrho\pi u_\infty^2 \sin\alpha \tag{7.18}$$

---

[2] From a theoretical point of view, this is hardly surprising, as torques already occur in the flow around an ellipse (Sect. 4.6).

Fig. 7.4 The wing cross-section is modelled as an infinitely thin plate. In the flow it receives a torque in addition to the lift.

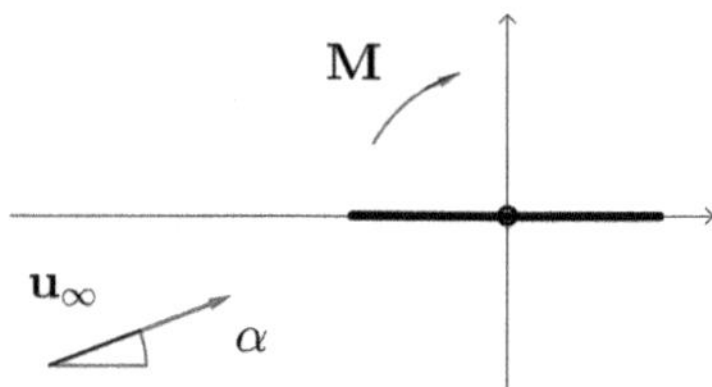

Proof From (7.13) we obtain

$$\left(\frac{dw}{d\zeta}\right)^2 = \left(u_\infty\left[e^{-i\alpha} - \frac{e^{i\alpha}}{\zeta^2}\right] - \frac{i\Gamma}{2\pi}\frac{1}{\zeta}\right)^2$$

Substituting into (S.9) yields

$$z\left(\frac{d\hat{w}}{dz}\right)^2 dz = \underbrace{\left(u_\infty\left[e^{-i\alpha} - \frac{e^{i\alpha}}{\zeta^2}\right] - \frac{i\Gamma}{2\pi}\frac{1}{\zeta}\right)^2}_{A} \underbrace{\left(\zeta + \frac{2}{\zeta} + \frac{2}{\zeta^3} + \dots\right)}_{B} d\zeta \qquad (7.19)$$

We evaluate (7.16), (7.19) by formula (A.21) so that only powers $1/\zeta$ contribute to the integrand. These arise either through multiplication of terms $\sim 1/\zeta^2$ from A by terms $\sim \zeta$ from B, or through multiplication of terms $\sim \zeta^0$ from A by terms $\sim 1/\zeta$ from B. If we denote by O the part of AB that is not of order $O(1/\zeta)$, we obtain

$$z\left(\frac{d\hat{w}}{dz}\right)^2 dz = \left(\left[-2u_\infty^2 e^{-i\alpha}\frac{e^{i\alpha}}{\zeta^2} - \frac{\Gamma^2}{4\pi^2}\frac{1}{\zeta^2}\right]\zeta + u_\infty^2 e^{-2i\alpha}\frac{2}{\zeta} + O\right)d\zeta$$

$$= \left(\left[-2u_\infty^2 - \frac{\Gamma^2}{4\pi^2} + 2u_\infty^2(\cos(2\alpha) - i\sin(2\alpha))\right]\frac{1}{\zeta} + O\right)d\zeta$$

Taking into account (A.21) we get

$$\int_C z\left(\frac{d\hat{w}}{dz}\right)^2 dz = \left[-2u_\infty^2 - \frac{\Gamma^2}{4\pi^2} + 2u_\infty^2\cos(2\alpha) - 2u_\infty^2 i\sin(2\alpha)\right]2\pi i$$

$$\mathrm{Re}\left[\int_C z\left(\frac{d\hat{w}}{dz}\right)^2 dz\right] = \left[-2u_\infty^2 i\sin(2\alpha)\right]2\pi i = 4u_\infty^2\sin(2\alpha)\pi$$

Because $\alpha \approx 0$, it holds $\cos\alpha \approx 1$ and thus $\sin(2\alpha) = 2\sin\alpha\cos\alpha \approx 2\sin\alpha$. Therefore, we find that

$$M = -\frac{\varrho}{2}\mathrm{Re}\left[\int_C z\left(\frac{d\hat{w}}{dz}\right)^2 dz\right] = -4\varrho\pi u_\infty^2\sin\alpha$$

The negative sign indicates the clockwise direction of rotation.   □

Fig. 7.5 The Fokker D.VIII monoplane was deployed from July 1918. After initial design flaws were eliminated, it proved to be very effective, but could no longer influence the course of the war.

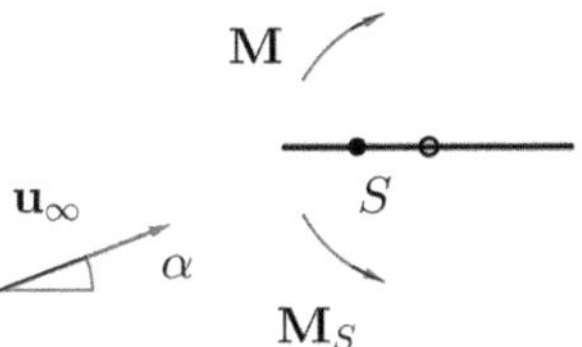

Fig. 7.6 If the centre of gravity S is to the left of the midpoint, a compensating torque M_S is created. In modern aircraft, small amounts of depleted uranium, which has a high density, are installed to achieve final weight balance.

When exiting the dive, the required increase in speed in combination with high angles of attack led to increased torque. This tore apart the lightweight wooden construction of the monoplane (see Fig. 7.5), which had been softened by infiltrated condensation water. Gravity and lift alone could not be the reason, since the wings could withstand six times the weight of the aircraft. Finally, wind tunnel measurements showed a lifting of the leading edge, indicating the effect of a torque.

To ensure stability, the torque must be balanced by a mass distribution that shifts the centre of gravity to the left (Fig. 7.6). Through several improvements, it was possible to stabilise the airplane without significant weight increase. Afterwards, it proved to be clearly superior to enemy aircraft (see [3]). The era of the monoplane had begun.

7.3   Flight Characteristics of Aircraft

Now we examine some problems of energy consumption, performance and manoeuvrability of aircraft. The following key figure proves to be useful.

Fig. 7.7 Forces in straight
flight.

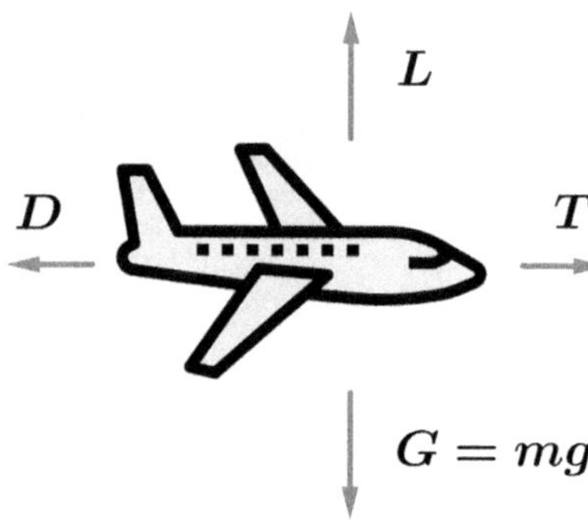

The *wing loading* is defined as the ratio

$$W_s = \frac{mg}{S},$$
(7.20)

where S is the wing area.

Instead of the previous wing-fixed coordinate system, we choose the coordinate
origin at a fixed point in space. Thus, u_∞ is the flight speed.

Straight and Steady Flight

A *straight and steady flight* of an aircraft means that it flies at constant
velocity.

In straight-line flight, the following forces act (Fig. 7.7)

$$
\begin{array}{ll}
T & -\,Thrust \\
D & -\,Drag \\
L & -\,Lift \\
G = mg & -\,Weight
\end{array}
$$

To maintain flight altitude, lift and weight must be equal, i.e.

$$L = mg \qquad \text{(straight flight at constant altitude)}$$
(7.21)

If the aircraft also flies at a constant speed, then

$$T = D \tag{7.22}$$

Minimum Required Power and Maximum Time Aloft

In the event of short-term weather deterioration at the destination, technical problems during landing and other situations, it may be advisable to keep an aircraft in the air for longer. We are looking for the most energy-efficient flight speed for straight and steady flight, neglecting the mass loss due to fuel combustion. Fuel consumption is lowest when the aircraft requires the minimum power to maintain its altitude. With the notations

$$
\begin{array}{ll}
\Delta s & - \text{distance travelled,} \\
\Delta t & - \text{required flight time,} \\
u_\infty = \Delta s/\Delta t & - \text{flight speed,} \\
W & - \text{work done for propulsion,} \\
P & - \text{power,}
\end{array}
$$

it holds for a uniform linear motion that

$$P = \frac{W}{\Delta t} = \frac{T\,\Delta s}{\Delta t} = Tu_\infty$$

and by (7.22)

$$P = Du_\infty \tag{7.23}$$

Problem 7.4 Prove for the drag in straight and steady flight that

$$D = \frac{1}{2}\varrho_\infty u_\infty^2 SC_d + \frac{2m^2g^2}{\varrho_\infty u_\infty^2 S}\frac{1}{\pi e\!\cancel{R}}, \tag{7.24}$$

where e denotes the Oswald efficiency number.

If we substitute (7.24) into (7.23), we get the required power (see Fig. 7.8)

$$P = \frac{1}{2}\varrho_\infty u_\infty^3 SC_d + \frac{2m^2g^2}{\varrho_\infty u_\infty S}\frac{1}{\pi e\!\cancel{R}} \tag{7.25}$$

Theorem 7.2 *In straight and steady flight, the power P is minimised at speed*

$$u_{minP} = \left[\frac{1}{3}\left(\frac{2W_s}{\varrho_\infty}\right)^2\frac{1}{C_d\,\pi e\!\cancel{R}}\right]^{1/4} \tag{7.26}$$

This is the speed at which the maximum time aloft is achieved.

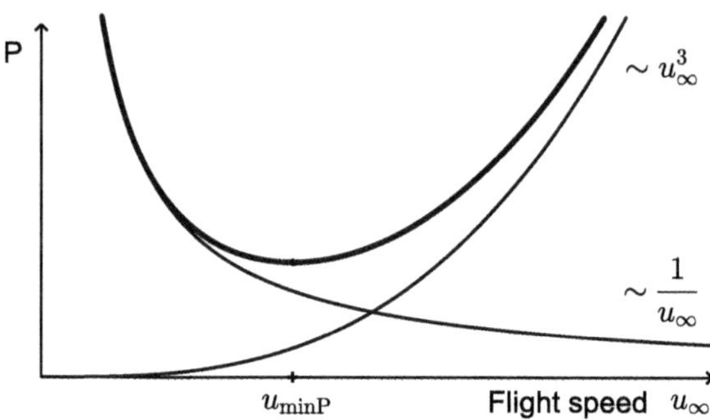

Fig. 7.8 Power required to maintain flight altitude as a function of speed.

Proof From (7.25) it follows that

$$\frac{\mathrm{d}P}{\mathrm{d}u_\infty} = \frac{3}{2}\varrho_\infty S C_d u_\infty^2 - \frac{2m^2 g^2}{\varrho_\infty S}\frac{1}{\pi e \mathcal{R}}u_\infty^{-2}$$

$$\frac{\mathrm{d}^2 P}{\mathrm{d}u_\infty^2} = 3\varrho_\infty S C_d u_\infty + \frac{4m^2 g^2}{\varrho_\infty S}\frac{1}{\pi e \mathcal{R}}u_\infty^{-3} > 0$$

The minimum of P is found under the condition $\mathrm{d}P/\mathrm{d}u_\infty = 0$. If we replace S by its expression from (7.20), we get the desired formula (7.26). □

Minimum Drag Losses and Maximum Range

The drag D should be kept small, as it reduces the thrust T and does not contribute to lift. We are looking for the flight speed to minimise it.

Problem 7.5 Prove the following formula for straight and steady flight.

$$D = mg\left(\frac{C_d}{C_L} + \frac{C_L}{\pi e \mathcal{R}}\right) \tag{7.27}$$

Theorem 7.3 *For straight and steady flight, the drag D is minimised at the speed*

$$u_{minD} = \left[\left(\frac{2W_s}{\varrho_\infty}\right)^2 \frac{1}{C_d\,\pi e \mathcal{R}}\right]^{1/4} \tag{7.28}$$

and the lift coefficient

$$C_{L,minD} = \sqrt{C_d\,\pi e \mathcal{R}} \tag{7.29}$$

Proof From (7.27) it follows that

$$\frac{\mathrm{d}D}{\mathrm{d}C_L} = mg\left(-\frac{C_d}{C_L^2} + \frac{1}{\pi e \mathcal{R}}\right), \qquad \frac{\mathrm{d}^2 D}{\mathrm{d}C_L^2} = mg\,\frac{2C_d}{C_L^3} > 0$$

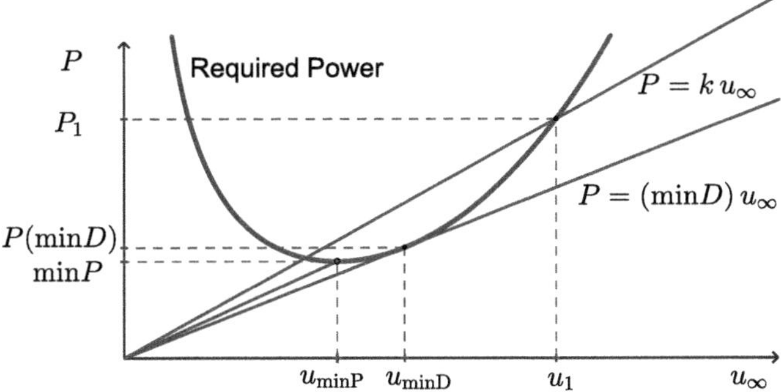

Fig. 7.9 The flight speed u_{minD} is determined by the line with the smallest slope, which connects the origin of the coordinates with a point on the power curve.

Applying the condition $dD/dC_L = 0$ for the minimum of D, we obtain (7.29). By rearranging (6.16), we find that

$$u_\infty = \left(\frac{L}{\frac{1}{2}\varrho_\infty S C_L} \right)^{1/2}$$

If we replace L by (7.21) and C_L by (7.29), we get (7.28). $\square$

With minimal drag losses, the engine power can be optimally used for propulsion. Therefore, the maximum range is achieved at the speed u_{minD}. Comparing (7.26) with (7.28) yields

$$u_{minP} = \frac{1}{\sqrt[4]{3}} \, u_{minD} \approx 0.76 \, u_{minD},$$

i.e., the speed at maximum time aloft is 76% of the speed for maximum range.

The power-speed diagram (Fig. 7.9) explains how u_{minD} can be determined graphically. We consider a line $P = k \, u_\infty$, which intersects the power curve (7.25) at a certain point (u_1, P_1). Owing to (7.23), we have $P_1 = D \, u_1$, i.e. the values of flight speed and required power are connected by a straight line with slope D. The smallest possible slope is the minimum value of D. This is the case when the line is tangent to the power curve.

For simplicity, we assumed that the weight is constant. In [4], the mass loss due to fuel combustion is taken into account. This allows not only speeds, but also the maximum time aloft or maximum range to be calculated.

Calculation of the Turning Radius

We examine maneuverability using an example from World War II. (see [1]).

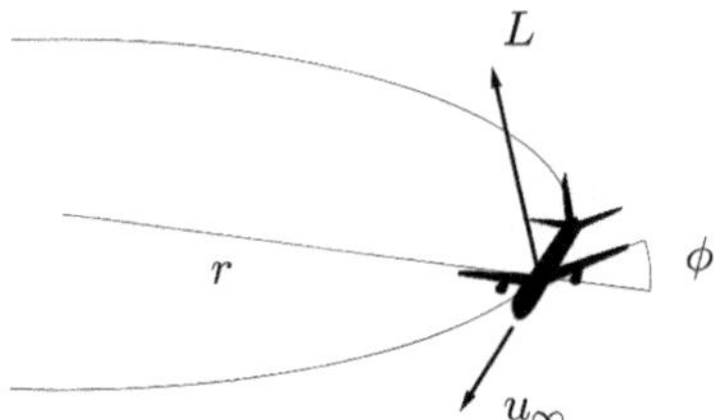

Fig. 7.10 Bank angle, lift and radius during the turning manoeuvre.

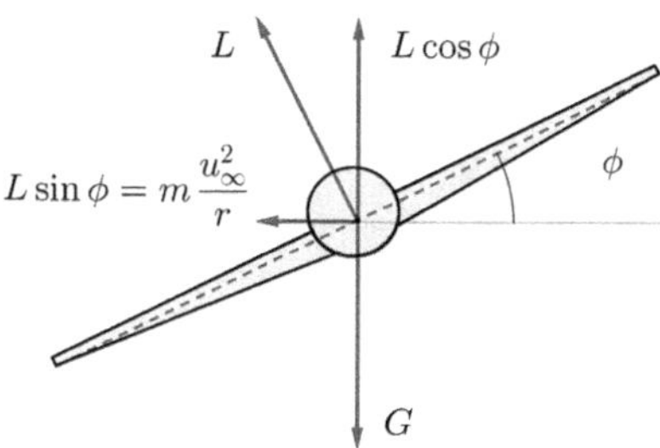

Fig. 7.11 Front view with force balance.

To fly a circle in the horizontal plane, the aircraft must tilt relative to the horizontal. Since the lift L always acts perpendicular to the wing plane, it tilts at the same angle. Now it can be decomposed into two components, with the horizontal component enabling the turning maneuver (see Figs. 7.10 and 7.11).

> The *bank angle* ϕ is the angle of inclination of the chord line relative to the horizontal.
>
> The radius r of the turning circle is referred to as the *turning radius*.

The horizontal component of the lift provides the centripetal force for the circular motion.

$$L \sin \phi = m \frac{u_\infty^2}{r} \tag{7.30}$$

By (6.16) we get the lift.

$$L = \frac{1}{2} \varrho_\infty u_\infty^2 S C_l$$

Using (7.30) and the wing loading $W_s = mg/S$, we obtain the turning radius.

$$r = \frac{2 W_s}{\varrho_\infty g C_L \sin \phi} \tag{7.31}$$

Fig. 7.12 The confiscated Bf 109E-3 after its transfer to Wright Field, Ohio (USA). Photo: USAAF.

Fig. 7.13 Supermarine Spitfire K9795. The elliptical wings reduce the wing loading W_s by about a quarter compared to the Bf 109. Photo: RAF.

In addition to a larger lift coefficient C_L, a lower wing loading contributes to manoeuvrability. With increasing altitude, the radius increases because the air density ϱ decreases.

The banking manoeuvre reduces the lift, which is now only provided by the vertical component $L \cos\phi$ instead of L. If there was a force balance $L = mg$ in straight flight, the tilting of the wings creates a downward force

$$F = mg - L \cos\phi$$

and the pilot must stabilise the aircraft to prevent a loss of altitude. According to (4.64) it is possible to increase the angle of attack, but a critical value must not be exceeded (Fig. 6.8). By (6.53), it must also be taken into account that with increasing lift C_L, the induced drag C_{D_i} also increases. The engine requires sufficient power to maintain the flight speed.

Example 7.1 During the Battle of Britain between 1940 and 1941, the Germans deployed the Messerschmitt Bf 109E and Bf 110 fighter aircraft (Fig. 7.12), against which the British countered with the outdated Boulton-Paul Defiant, the robust and safe Hawker Hurricane, and the promising newly developed Supermarine Spitfire (Fig. 7.13). During the previous battles in 1939, an intact Bf 109E-3 had fallen into the hands of the French, who handed it over to the British for evaluation. The list of points of interest included a comparison of their manoeuvrability with the Spitfire. Table 7.1 contains results calculated from the measured values.

Table 7.1 Comparison data of Messerschmitt Bf 109E-3 and Supermarine Spitfire. The evaluation was carried out at an altitude of 12,000 ft (3658 m). For C_L, values at the maximum were chosen. Data source: [6].

	ϕ (degrees)	C_L	u_∞ (km/h)	radius r (m)
Spitfire	68	1.47	203	212
Bf 109	62	1.60	190	270

They confirm that the Spitfire was more maneuverable and faster despite its lower lift coefficient. The elliptical wings contributed to these advantages (see [1]).

References and Further Reading

1. Ackroyd J.A.D.: *The Aerodynamics of the Spitfire.* (2017) https://api.semanticscholar.org/CorpusID:199554973 Accessed 3rd Sept 2024
2. Fabian S.T., Zhou R., Lin H-T.: *Dragondrop: a novel passive mechanism for aerial righting in the dragonfly.* Proc. R. Soc. (2021), B 288: 20202676.
3. Gordon J.E.: *Structures Or Why Things Don't Fall Down.* DeCapo Press. 2nd Edition (2003)
4. Marchman J.F.: *Aerodynamics and Aircraft Performance.* Virginia Tech (2022)
5. McNeill Alexander R.: *Principles of Animal Locomotion.* Princeton University Press, Princeton (2003)
6. Morgan M.B., Morris D.E.: *Messerschmitt Me 109. Handling and manoeuvrability tests.* ARC, R & M No. 2361 (1940)
7. Oertel H.jr., Ruck S.: *Bioströmungsmechanik.* Vieweg+Teubner, Wiesbaden (2012)

Some Applications from Hydrodynamics 8

In special cases, the Navier-Stokes equations can be reduced to a solvable form by neglecting insignificant components. The presented examples illuminate a number of remarkable natural phenomena.

8.1 Gravity Waves

The impressive waves on the open sea are called *gravity waves*. They wash the shores of islands from all sides, which is by no means self-evident for originally straight wave fronts.

When swimming in strong swell (and strict bathing prohibition), one notices that the particles in the wave perform a circular motion. The "washing machine" puts inexperienced people into panic and is regularly responsible for bathing accidents. Experienced swimmers dive immediately before strong waves, as the intensity of the rotational movement decreases towards the bottom (Fig. 8.4).

Vortex-free Model

Our problem is complicated by the fact that only the lower boundary of the region is fixed. The upper boundary, i.e. the water/air interface, is not only unknown but also time-dependent. This creates an additional degree of freedom. Here we speak of a problem with a *free surface*. We are looking for a *non-stationary velocity field* **u**, which fulfils the following *Airy conditions*.

R. Spielmann, *Theoretical Fluid Mechanics*,
https://doi.org/10.1007/978-3-662-72240-4_8

Fig. 8.1 Boundary
conditions.

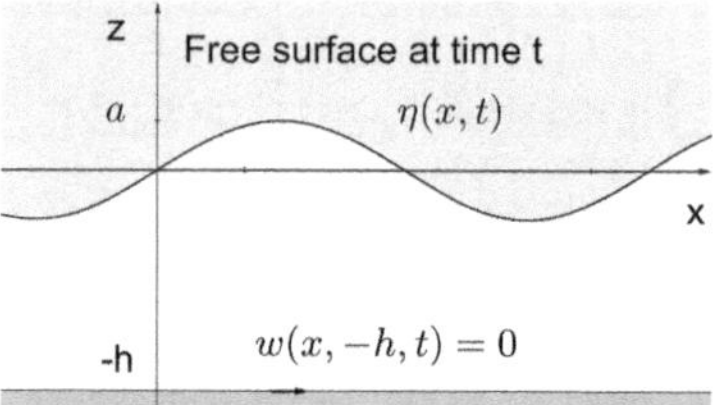

Fig. 8.2 Change of the free
surface after half a period.

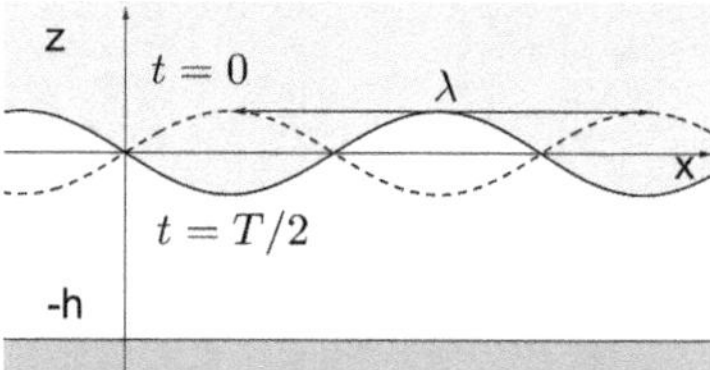

A1 The density ϱ is constant. Our fluid is both *incompressible* and *vortex-free*. Thus, there is a potential flow $\mathbf{u} = \nabla\phi_1$.

A2 At a sufficient distance from the shore, the wave does not break.
  Observations confirm that a particle from the free surface remains there for all times. We describe the upper boundary of the flow region by trajectories (see Figs. 8.1 and 8.2)

$$z = \eta(x,t) \quad \text{with } p = p_{\text{atm}} \text{ (atmospheric air pressure)} \tag{8.1}$$

In our coordinate system, the x-axis points in the direction of propagation and the z-axis points upwards. The equilibrium position of the free surface is at $z = 0$ and the ground at $z = -h$.

The basic building block of the modeling is the concept of a *monochromatic plane wave*. Here, the quantity a denotes the *amplitude*, T the *period*, λ the *wavelength*, and c the *phase velocity*. It holds

$$c = \frac{\lambda}{T} \tag{8.2}$$

Thus, we set

$$\mathbf{u} = \begin{pmatrix} u(x, z, t) \\ 0 \\ w(x, z, t) \end{pmatrix} = \begin{pmatrix} \frac{\partial \phi_1}{\partial x} \\ 0 \\ \frac{\partial \phi_1}{\partial z} \end{pmatrix}$$

From the non-stationary Bernoulli equation (4.8), it follows for $\Phi = gz$ and a time-dependent constant $c(t)$ that

$$\frac{\partial \phi_1}{\partial t} + \frac{p}{\varrho} + \frac{1}{2}(u^2 + w^2) + gz = c(t) \quad \text{in the entire flow region} \tag{8.3}$$

The constant $c(t)$ can be included in the potential if we use the approach

$$\phi_1 = \phi - \frac{p_{\text{atm}}}{\varrho} t + \int_{t_0}^{t} c(\tau)\, d\tau \tag{8.4}$$

to switch to the potential ϕ. Because

$$\nabla \phi_1 = \nabla \phi, \qquad \frac{\partial \phi_1}{\partial t} = \frac{\partial \phi}{\partial t} - \frac{p_{\text{atm}}}{\varrho} + c(t),$$

it follows from (8.3) that

$$\frac{p}{\varrho} - \frac{p_{\text{atm}}}{\varrho} + \frac{\partial \phi}{\partial t} + \frac{1}{2}(u^2 + w^2) + gz = 0 \tag{8.5}$$

The special case of (8.5) at the free surface $z = \eta(x, t)$ is referred to as the *dynamic boundary condition*

$$\frac{\partial \phi}{\partial t} + \frac{1}{2}(u^2 + w^2) + g\eta = 0 \tag{8.6}$$

To derive this, we set $z = \eta$ and $p = p_{\text{atm}}$.

Observations confirm that the vertical component of the particle velocity vanishes at the bottom, i.e.

$$w(x, -h, t) = \frac{\partial \phi}{\partial z}\Big|_{z=-h} = 0 \tag{8.7}$$

The *kinematic boundary condition* is defined as

$$\frac{\partial \eta}{\partial t} + u \frac{\partial \eta}{\partial x} = w \quad \text{along a surface trajectory} \tag{8.8}$$

Problem 8.1 Derive the kinematic boundary condition by taking the material derivative of the first equation of (8.1).

By A1 and (8.4), we get the *Laplace equation*

$$\Delta \phi = 0 \tag{8.9}$$

The Approximation of Linear Wave Theory

Since no complete solution of the system (8.6), (8.7), (8.8) and (8.9) is known, we are looking for an approximation. Scaling provides information about which terms dominate and which can be neglected.

- As a *length scale for the x-direction* we use the wavelength λ.
    For ocean waves, λ is typically on the order of 100 m.
- The *time scale* is the period T.
    The typical period for ocean waves is about 8 s.
- The *scale of the vertical displacement* at the water surface is the amplitude a. Therefore, we have $\eta = O(a)$.

From this, the remaining functions with their scales can be derived. For example, the velocity scale is determined by the vertical motion on the surface. In particular,

$$|v| = O(a/T)$$

For the dimensionless quantities a/λ and gT^2/λ, we make the following assumptions.

L1 $a \ll \lambda$ implies $a/\lambda \ll 1$. Therefore, terms of order $O(a/\lambda)$ can be neglected.
L2 Gravity is considered to be significant, i.e. we set $gT^2/\lambda = O(1)$.

Now we simplify the kinematic condition (8.8). Firstly,

$$\frac{\partial \eta}{\partial t} = O\left(\frac{a}{T}\right) \quad \text{and} \quad w = O\left(\frac{a}{T}\right)$$

Because

$$u \frac{\partial \eta}{\partial x} = O\left(\frac{a}{T}\frac{a}{\lambda}\right) = O\left(\frac{a}{T}\right) O\left(\frac{a}{\lambda}\right),$$

$u \frac{\partial \eta}{\partial x}$ is negligible compared to the terms $\frac{\partial \eta}{\partial t}$ and w. From (8.8) it follows that

$$\frac{\partial \eta}{\partial t} = w \tag{8.10}$$

Similarly, we simplify the dynamic condition (8.6). Since $\frac{\partial \phi}{\partial x} = O\left(\frac{a}{T}\right)$, we get $\phi = O\left(\frac{\lambda a}{T}\right)$ and therefore

$$\frac{\partial \phi}{\partial t} = O\left(\frac{\lambda a}{T^2}\right)$$

On the other hand,

$$u^2 + w^2 = O\left(\left(\frac{a}{T}\right)^2\right) = O\left(\frac{\lambda a}{T^2}\right) O\left(\frac{a}{\lambda}\right),$$

which is negligible compared to $\partial \phi/\partial t$. Finally, we examine the term $g\eta$. Since $gT^2/\lambda = O(1)$, we find that

$$g\eta = \frac{gT^2}{\lambda}\frac{\lambda \eta}{T^2} = O(1)\, O\left(\frac{\lambda a}{T^2}\right) = O\left(\frac{\lambda a}{T^2}\right),$$

i.e. $g\eta$ is of the same order as $\partial \phi/\partial t$. Thus we get from (8.6)

$$\frac{\partial \phi}{\partial t} + g\eta = 0 \quad \text{at the free surface } z = \eta(x, t) \tag{8.11}$$

Even the system (8.9), (8.7), (8.10), (8.11) is too complicated because the free surface is unknown. First, we simplify (8.10). Applying the mean value theorem, we obtain

$$w(x, \eta, t) = w(x, 0, t) + \frac{\partial w}{\partial z}\Big|_{z=\theta} \eta \qquad \text{for } \theta \in (0, \eta) \tag{8.12}$$

It holds $w(x, \eta, t) = O(a/T)$ and $w(x, 0, t) = O(a/T)$. Because of

$$\frac{\partial w}{\partial z}\Big|_{z=\theta} = O\left(\frac{a/T}{\lambda}\right), \qquad \frac{\partial w}{\partial z}\Big|_{z=\theta} \eta = O\left(\frac{a/T}{\lambda}a\right) = O\left(\frac{a}{T}\right) O\left(\frac{a}{\lambda}\right),$$

the term $\frac{\partial w}{\partial z}\big|_{z=0}\,\eta$ can be neglected in (8.12). Equation (8.10) simplifies to

$$\frac{\partial \eta}{\partial t} = w(x, 0, t) \tag{8.13}$$

In a similar way, we simplify (8.11) to

$$\frac{\partial \phi}{\partial t}\,\big|_{z=0} + g\eta = 0 \tag{8.14}$$

Solution of the Linearised System

We note the system (8.9), (8.7), (8.13), (8.14) again, replacing $w(x, 0, t)$ in (8.13) by $\frac{\partial \phi}{\partial z}\big|_{z=0}$.

$$\frac{\partial^2 \phi}{\partial x^2} + \frac{\partial^2 \phi}{\partial z^2} = 0 \qquad (-h \le z \le \eta) \tag{8.15}$$

$$\frac{\partial \phi}{\partial z}\,\big|_{z=-h} = 0 \tag{8.16}$$

$$\frac{\partial \eta}{\partial t} = \frac{\partial \phi}{\partial z}\,\big|_{z=0} \tag{8.17}$$

$$\frac{\partial \phi}{\partial t}\,\big|_{z=0} = -g\eta \tag{8.18}$$

Now we use the approach

$$\phi(x, z, t) = A(z)\,\sin(\omega t - kx + \phi_0) \tag{8.19}$$

which models the dependence of x and t in the form of a monochromatic, plane wave with parameters

$$\omega = \frac{2\pi}{T} \qquad \text{angular frequency} \tag{8.20}$$

$$k = \frac{2\pi}{\lambda} \qquad \text{wave number} \tag{8.21}$$

By (8.19) we obtain from (8.15)

$$(-k^2 A(z) + A''(z))\,\sin(\omega t - kx + \phi_0) = 0$$

Thus, $-k^2 A(z) + A''(z) = 0$ and consequently

$$A(z) = C_1 \cosh(kz + C_2) \tag{8.22}$$

We determine the constants C_1, C_2 and ϕ_0.

Problem 8.2 Use (8.16) to show that

$$C_2 = kh \tag{8.23}$$

By substitution into (8.22) we obtain the solution of (8.15), (8.16)

$$\phi(x, z, t) = C_1 \cosh(k(z + h)) \, \sin(\omega t - kx + \phi_0) \tag{8.24}$$

Let us now consider the Eqs. (8.17) and (8.18). From (8.18) it follows that

$$\eta = -\frac{1}{g} \frac{\partial \phi}{\partial t} \Big|_{z=0} = -\frac{\omega}{g} C_1 \cosh(kh) \, \cos(\omega t - kx + \phi_0) \tag{8.25}$$

$$\frac{\partial \eta}{\partial t} = \frac{\omega^2}{g} C_1 \cosh(kh) \, \sin(\omega t - kx + \phi_0) \tag{8.26}$$

Taking the derivative of (8.24) yields

$$\frac{\partial \phi}{\partial z} \Big|_{z=0} = k \, C_1 \sinh(kh) \, \sin(\omega t - kx + \phi_0) \tag{8.27}$$

By substitution of (8.26) and (8.27) into (8.17) we get

$$\frac{\omega^2}{g} C_1 \cosh(kh) \, \sin(\omega t - kx + \phi_0) = k \, C_1 \sinh(kh) \, \sin(\omega t - kx + \phi_0)$$

and thus the *dispersion relation*

$$\omega^2 = gk \tanh(kh) \tag{8.28}$$

By choosing the constants

$$\phi_0 = -\frac{\pi}{2}, \qquad C_1 = \frac{ag}{\omega \cosh(kh)}, \tag{8.29}$$

we obtain from (8.25)

$$\eta = a \sin(kx - \omega t).$$

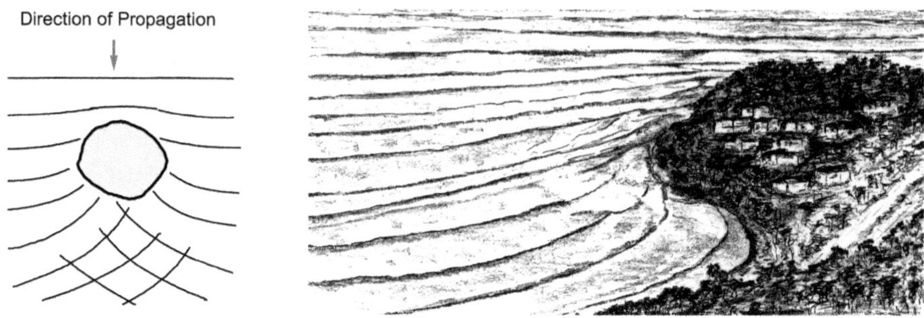

Fig. 8.3 (left) Wave fronts moving towards an island. As the wave reaches the shallow water, it slows down, causing it to be diffracted and to wash around the island. (right) Rincon, California.

So we have found the equation of the monochromatic plane wave to describe the free surface. Substituting (8.22), (8.23), (8.29) into (8.19), we get the potential

$$\phi(x, z, t) = \frac{ag}{\omega \cosh(kh)} \cosh(k(z+h)) \, \sin(\omega t - kx - \frac{\pi}{2})$$

$$= \frac{ag}{\omega} \frac{\cosh(k(z+h))}{\cosh(kh)} \, \cos(\omega t - kx) \tag{8.30}$$

Problem 8.3 Derive the following relation for the phase velocity.

$$c = \frac{1}{k}\sqrt{gk \tanh(hk)} \tag{8.31}$$

To illustrate, we examine the approximations for deep and shallow water. In deep water ($h \to \infty$) we have $\lim_{h \to \infty} \tanh(hk) = 1$ and therefore

$$c_{\text{deep}} = \sqrt{\frac{g}{k}} = \sqrt{\frac{g\lambda}{2\pi}} \tag{8.32}$$

For shallow water ($h \to 0$), it holds that $\lim_{h \to 0} \tanh(hk) = hk$, and thus

$$c_{\text{shallow}} = \frac{k\sqrt{gh}}{k} = \sqrt{gh} \tag{8.33}$$

As depth decreases, the propagation speed decreases, which explains why waves in shallow water are diffracted and as a consequence always run towards the shore (Fig. 8.3).

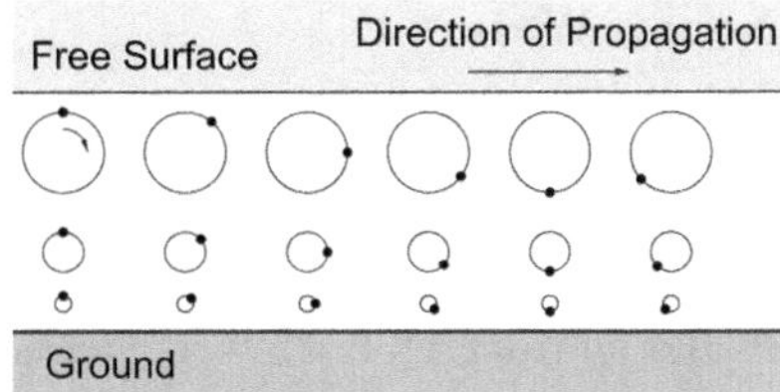

Fig. 8.4 Particles perform a circular motion. As a result, the water at the wave crest moves in the direction of the propagation and in the wave trough in the opposite direction. With increasing depth, the orbitals become exponentially smaller. This means that you are less exposed to rotation when diving.

To calculate the particle velocity, we simplify

$$\frac{\cosh(k(z+h))}{\cosh(kh)} = \frac{e^{k(z+h)} + e^{-k(z+h)}}{e^{kh} + e^{-kh}} = e^{kz}\frac{1 + e^{-2k(z+h)}}{1 + e^{-2kh}}$$

as well as relation (8.30) for deep water ($h \to \infty$)

$$\lim_{h\to\infty} \phi = \frac{ag}{\omega}\cos(\omega t - kx) \lim_{h\to\infty}\frac{\cosh(k(z+h))}{\cosh(kh)} = \frac{ag}{\omega}\cos(\omega t - kx)\,e^{kz}$$

$$= \frac{a\omega}{k}\cos(\omega t - kx)\,e^{kz} \qquad \text{using the dispersion relation } \omega = \frac{gk}{\omega}$$

This results in the horizontal and vertical components of the particle velocity

$$\mathbf{u} = \begin{pmatrix} u \\ 0 \\ w \end{pmatrix} = \begin{pmatrix} \frac{\partial\phi}{\partial x} \\ 0 \\ \frac{\partial\phi}{\partial z} \end{pmatrix} = \begin{pmatrix} a\omega e^{kz}\sin(\omega t - kx) \\ 0 \\ a\omega e^{kz}\cos(\omega t - kx) \end{pmatrix}$$

The formula shows that the particles perform a circular motion whose amplitude decreases exponentially (Fig. 8.4), i.e. the sea becomes calmer with increasing depth. At the wave crest, the orbital motion runs in the direction of propagation, while it is opposite in the wave trough. In the ocean, this phenomenon can cause unpleasant surprises for unsuspecting bathers. In rough seas, a swimmer is literally whirled around, which quickly leads to panic and often ends in drowning.

In the shore surf, the upper circular movement is greater than in the trough and the wave breaks. This is associated with greater mass transportation.

8.2   Currents in Saline Waters

During the Second World War, several German submarines managed to pass from the Atlantic through the narrow, carefully guarded Strait of Gibraltar into the Mediterranean, where they remained for the rest of the war. The risky dive was facilitated by an underwater current that allowed stealthy travel with minimal engine power. We show that such currents are typical for passages between regions of water with different salinity.

With increasing salinity, the density of seawater increases. As it turns out, this creates currents that flow at the surface towards the saltier region and in the depth towards the less salty region. In the following, we assume that pressure and temperature in different regions are identical, so that the density is determined only by the salinity. On a closed curve C moving with the flow in the yz-plane (see Fig. 8.5), we consider the particle acceleration $\mathbf{a} = d\mathbf{u}/dt$ as well as

$$\mathbf{B} = \operatorname{rot} \mathbf{a}$$

By Σ we denote the area enclosed by C. Applying Stokes' theorem (Theorem 1.2), we find that

$$\oint_C \mathbf{a}\, d\mathbf{x} = \int_\Sigma (\operatorname{rot} \mathbf{a}) \cdot \mathbf{n}\, d\Sigma = \int_\Sigma \mathbf{B} \cdot \mathbf{n}\, d\Sigma$$

Let us show that this integral has a positive value. Owing to (2.4), (3.30), we find for a inviscid fluid in a conservative force field with force density $\mathbf{f} = \nabla \omega$, that

$$\mathbf{a} = -\frac{1}{\varrho}\nabla p - \nabla \omega$$

Here, ω denotes the potential density from Sect. 1.3. Applying the curl, we obtain

$$\mathbf{B} = -\operatorname{rot}\left(\frac{1}{\varrho}\nabla p\right) \qquad \text{by (1.19)}$$

$$= \nabla p \times \nabla\left(\frac{1}{\varrho}\right) \qquad \text{by (1.13)}$$

The orientation of $\mathbf{B}$ results from the following properties of p and $1/\varrho$.

- The pressure increases downwards. Thus, $\nabla p = c_1 \begin{pmatrix} 0 \\ 0 \\ -1 \end{pmatrix}$ for $c_1 > 0$.

- The density decreases in the direction of the less salty water. According to Fig. 8.5, we have

$$\nabla\left(\frac{1}{\varrho}\right) = c_2 \begin{pmatrix} 0 \\ 1 \\ 0 \end{pmatrix} \qquad \text{for } c_2 > 0$$

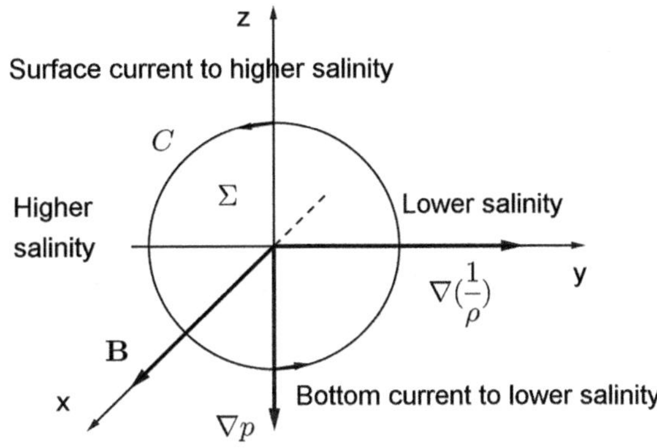

Fig. 8.5 Between bodies of water with different salinity, currents form in different directions.

Therefore,

$$\mathbf{B} = \nabla p \times \nabla(\frac{1}{\varrho}) = c_1 c_2 \begin{pmatrix} 1 \\ 0 \\ 0 \end{pmatrix}$$

If C is traversed in the mathematically positive sense, the surface normal $\mathbf{n}$ points in the same direction as $\mathbf{B}$. From $\mathbf{B} \cdot \mathbf{n} = c_1 c_2 > 0$ it follows that

$$\oint_C \mathbf{a}\,d\mathbf{r} = \int_\Sigma \mathbf{B} \cdot \mathbf{n}\,d\Sigma > 0$$

Since $\mathbf{a} \cdot d\mathbf{r} > 0$, the acceleration vector $\mathbf{a} = d\mathbf{u}/dt$ is also oriented in the direction of C. This results in the flow conditions shown in Fig. 8.5.

Different salt concentrations can be found between the Atlantic (A) and the Mediterranean (M) at the Strait of Gibraltar ($\varrho_A > \varrho_M$) as well as between the Mediterranean (M) and the Black Sea (S) at the Bosporus ($\varrho_M > \varrho_S$). There are also significant differences in salinity within the Baltic Sea, as its remote northeastern bays are fed by freshwater inflows. However, the significance of our model is reduced because different wind directions, water temperatures and tides must also be taken into account. Thus, the currents at the Bosporus are more pronounced, while they are partially overlaid by other influences at the Strait of Gibraltar (see [1], [2]).

Problem 8.4

(a) Let C be a closed curve moving with the flow in the yz-plane. Show that

$$\oint_C \mathbf{u}\,d\mathbf{u} = 0$$

(b) Prove that the rate of change of circulation can be expressed in terms of particle acceleration $\mathbf{a} = d\mathbf{u}/dt$.

$$\frac{d\Gamma}{dt} = \oint_C \mathbf{a}\,d\mathbf{x}$$

8.3   The Law of Hagen-Poiseuille

The following law explains why the autonomic nervous system enables efficient blood pressure regulation. It also becomes clear how drastically pathological vascular constrictions affect this.

We consider the unaccelerated flow through a vertically standing, circular cylinder without the influence of gravity. The calculations are based on formulas in cylindrical coordinates, which are derived in Sect. A.8.

The *volumetric flow rate* dV/dt of a fluid is defined as the volume of fluid that is passing through a given cross-sectional area per unit time.

Theorem 8.1 (Hagen-Poiseuille)   *We consider the volumetric flow rate dV/dt of a gas or a liquid in a long pipe with a radius a. Then we have*

$$\frac{dV}{dt} = -\frac{\pi a^4}{8\mu}\frac{\partial p}{\partial z} \qquad \text{in a short section of the pipe.}$$

Here, $\partial p/\partial z$ denotes the pressure drop in the direction of flow and $\mu = \varrho \nu$ the dynamic viscosity.

Proof We apply the Navier-Stokes equations in cylindrical coordinates (see Sect. A.8). The fluid flows parallel to the pipe axis, with its velocity at a point depending only on the distance to the axis. We use the approach

$$\mathbf{u} = \begin{pmatrix} u_r(r, \theta, z) \\ u_\theta(r, \theta, z) \\ u_z(r, \theta, z) \end{pmatrix} = \begin{pmatrix} 0 \\ 0 \\ u_z(r) \end{pmatrix}, \tag{8.34}$$

so that of the Eqs. (A.55), (A.56), (A.57) only (A.57) remains and we get

$$\frac{Du_z}{Dt} + \frac{1}{\varrho}\frac{\partial p}{\partial z} = \nu \Delta u_z \tag{8.35}$$

with

$$\frac{Du_z}{Dt} = \frac{\partial u_z}{\partial t} + u_r \frac{\partial u_z}{\partial r} + \frac{u_\theta}{r}\frac{\partial u_z}{\partial \theta} + u_z \frac{\partial u_z}{\partial z} \qquad \text{by (A.54)}$$

$$\Delta u_z = \frac{\partial^2 u_z}{\partial r^2} + \frac{1}{r}\frac{\partial \phi}{\partial r} + \frac{1}{r^2}\frac{\partial^2 u_z}{\partial \theta^2} + \frac{\partial^2 u_z}{\partial z^2} \qquad \text{by (A.50)}$$

From (8.34) we obtain

$$\frac{Du_z}{Dt} = 0$$

$$\Delta u_z = \frac{\partial^2 u_z}{\partial r^2} + \frac{1}{r}\frac{\partial u_z}{\partial r} = \frac{1}{r}\frac{\partial}{\partial r}\left(r\frac{\partial u_z}{\partial r}\right)$$

If we replace $\nu = \mu/\varrho$, then (8.35) simplifies to

$$\frac{r}{\mu}\frac{\partial p}{\partial z} = \frac{\partial}{\partial r}\left(r\frac{\partial u_z}{\partial r}\right)$$

By integrating twice (with the integration constants A,B), we get

$$\frac{r^2}{2\mu}\frac{\partial p}{\partial z} + A = r\frac{\partial u_z}{\partial r} \qquad \text{as a result of the first integration,}$$

$$\frac{r}{2\mu}\frac{\partial p}{\partial z} + \frac{A}{r} = \frac{\partial u_z}{\partial r} \qquad \text{after rearrangement,}$$

$$\frac{r^2}{4\mu}\frac{\partial p}{\partial z} + A \ln r + B = u_z \qquad \text{as a result of the second integration}$$

Fig. 8.6 The quantity flowing through depends on the radius of the layer.

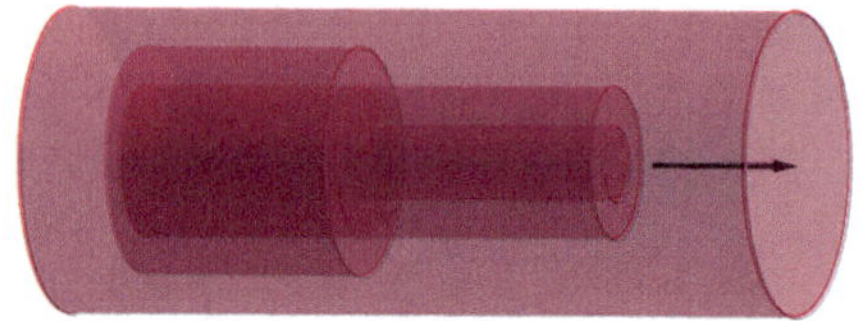

Because $\lim_{r\to 0} u_z(r) = 0$ and $\lim_{r\to\infty} \ln r = -\infty$, it holds $A = 0$. This gives

$$u_z = \frac{r^2}{4\mu}\frac{\partial p}{\partial z} + B$$

From the boundary condition $u_z(a) = 0$, we get $B = -\frac{a^2}{4\mu}\frac{\partial p}{\partial z}$ and therefore

$$u_z = \frac{r^2 - a^2}{4\mu}\frac{\partial p}{\partial z}$$

The maximum speed is reached on the central axis ($r = 0$): $u_{z,\max} = -\frac{a^2}{4\mu}\frac{\partial p}{\partial z}$.

For the volumetric flow rate, we add the volumes of the nested hollow cylinders of thickness dr, whose height increases inwards (see Fig. 8.6). Thus,

$$dV = \int_0^a \underbrace{2\pi r\, dr}_{\text{annular area}} \cdot \underbrace{u_z\, dt}_{\text{height}}$$

$$dV = \frac{\pi\, dt}{2\mu}\frac{\partial p}{\partial z}\int_0^a (r^3 - a^2 r)\, dr = -\frac{\pi\, dt}{2\mu}\frac{\partial p}{\partial z}\frac{a^4}{4}$$

$$\frac{dV}{dt} = -\frac{\pi a^4}{8\mu}\frac{\partial p}{\partial z}$$

□

In applications, we consider pipes in which a pressure drop Δp occurs along the pipe length l. Hence $\frac{\partial p}{\partial z} = \frac{\Delta p}{l}$ and therefore

$$\frac{dV}{dt} = -\frac{\pi a^4 \Delta p}{8\mu l} \tag{8.36}$$

The law can be applied to the vessels of the circulatory system and the lungs.

Problem 8.5   Under constant pressure conditions, the radius should be reduced by 20%. Show that the volumetric flow rate decreases by 60%.

The autonomic nervous system regulates blood supply by simultaneously causing a change in the pumping capacity of the heart muscle and the contraction or expansion of blood vessels. The latter are largely surrounded by a thin layer of smooth muscle. According to (8.36), the dependence in the fourth power ensures that minimal changes in radius significantly change the volumetric flow rate. This allows vascular flow to be regulated even by weaker muscles, and the Hagen-Poiseuille law explains the efficiency with which the autonomic nervous system enables adapted blood circulation. At the same time, it becomes obvious what danger threatens in the case of pathological vessel constriction.

References and Further Reading

1. Gonzalez C.J., Reyes E., Alvarez O., Izquierdo A., Bruno M., Mananes R.: *Surface currents and transport processes in the Strait of Gibraltar: Implications for modeling and management of pollutant spills.* Ocean & Coastal Management, Vol. 179 (2019)
2. Gregg, M.C., Özsoy E.: *Flow, water mass changes, and hydraulics in the Bosphorus.* J. Geophys. Res., 107(C3), (2002).
3. Milne-Thomson L.M.: *Theoretical Hydrodynamics.* Dover Publications, New York (1968)
4. Oertel H.jr., Ruck S.: *Bioströmungsmechanik.* Vieweg+Teubner, Wiesbaden (2012)
5. Trautwein A.X., Kreibig U., Hüttermann J.: *Physik für Mediziner, Biologen, Pharmazeuten.* De Gruyter Studium, Berlin (2014)

Correction to: Theoretical Fluid Mechanics

**Correction to:
R. Spielmann, *Theoretical Fluid Mechanics*,
https://doi.org/10.1007/978-3-662-72240-4**

The original version of the book was inadvertently published with errors in Chapters 6 and 7.

- In Chapter 6: Lifting Line Theory: On page 134, last line the text '(see Fig. 6.5) and Problem 6.6)' should read as '(see Fig. 6.4) and Problem 6.6'.

- In Chapter 7: Special Aerodynamic Problems:

  1. In page 164 the block

$$
\begin{aligned}
T &\ \text{----}\ \textit{Thrust}\\
D &\ \text{----}\ \textit{Drag}\\
L &\ \text{----}\ \textit{Lift}\\
G = mg &\ \text{----}\ \textit{Weight}
\end{aligned}
$$

  should read as

$$
\begin{aligned}
T &\quad -\textit{Thrust}\\
D &\quad -\textit{Drag}\\
L &\quad -\textit{Lift}\\
G = mg &\quad -\textit{Weight}
\end{aligned}
$$

The updated versions of these chapters can be found at
https://doi.org/10.1007/978-3-662-72240-4_6
https://doi.org/10.1007/978-3-662-72240-4_7

2. In page 165 the block

$$\Delta s \quad ----- \quad \text{distance travelled,}$$
$$\Delta t \quad ----- \quad \text{required flight time,}$$
$$u_\infty = \Delta s / \Delta t \quad ----- \quad \text{flight speed,}$$
$$W \quad ----- \quad \text{work done for propulsion,}$$
$$P \quad ----- \quad \text{power,}$$

should read as

$$\Delta s \qquad\qquad - \text{distance travelled,}$$
$$\Delta t \qquad\qquad - \text{required flight time,}$$
$$u_\infty = \Delta s / \Delta t \quad - \text{flight speed,}$$
$$W \qquad\qquad - \text{work done for propulsion,}$$
$$P \qquad\qquad - \text{power,}$$

3. In page 169 on the 11th line from the bottom, the text 'exceeded (Fig. 6.9)' should read as 'exceeded (Fig. 6.8)'.

A.1   Order of a Function

We call the functions f and g
- on the set E of *the same order*, for short $f = O(g)$ on E, if there exists a constant C such that

$$|f(x)| \le C\,|g(x)| \qquad \text{for } x \in E$$

- for $x \to a$ of *the same order*, for short $f = O(g)$ for $x \to a$, if in a certain neighbourhood $U(a)$ applies

$$f(x) = O(g(x)) \qquad \text{for } x \in U(a),\ x \ne a$$

In particular, we write

$$f(x) = O(1) \qquad \text{on } E$$

if f is bounded on E. The notation

$$f(x) = g(x) + O(h(x)) \qquad \text{for } x \to a$$

signify that $f(x) - g(x) = O(h(x))$ for $x \to a$.

© The Author(s), under exclusive license to Springer-Verlag GmbH, DE,
part of Springer Nature 2026
R. Spielmann, *Theoretical Fluid Mechanics*,
https://doi.org/10.1007/978-3-662-72240-4

A.2 Rotations and Isotropic Tensors

First, we repeat the index rules for matrix multiplication. We calculate the product of the matrices $A = (a_{ij})$ and $B = (b_{jk})$ by

$$AB = (a_{ij}b_{jk}) \quad \text{(summation over inner indices)} \tag{A.1}$$

Accordingly, we get for the *transposes* $A^T = (a_{ji})$, $B^T = (b_{kj})$

$$A^T B = (a_{ji}b_{jk}) \quad \text{(summation over first indices)} \tag{A.2}$$

$$AB^T = (a_{ij}b_{kj}) \quad \text{(summation over second indices)} \tag{A.3}$$

A *special orthogonal matrix* is a square matrix with a positive determinant, which satisfies the following conditions:

$$L^T L = LL^T = I \tag{A.4}$$

$$\text{or} \quad l_{ki}l_{kj} = l_{ik}l_{jk} = \delta_{ij}$$

Such matrices describe rotations of the coordinate system by allowing the calculation of the new coordinates $(\bar{x}_i)$ from the old coordinates (x_i) for each point in space.

$$\bar{x}_j = l_{ij}x_i \quad \text{(summation over the first index)} \tag{A.5}$$

Problem A.1 Prove the formula

$$x_i = l_{ij}\bar{x}_j \quad \text{(summation over the second index)} \tag{A.6}$$

for the inverse of the transformation (A.5).

A *tensor* A of rank n is an object with 3^n components $a_{i_1...i_n}$ that can be transformed into a new coordinate system by rotation as follows.

$$\bar{a}_{j_1...j_n} = l_{i_1 j_1} \ldots l_{i_n j_n} a_{i_1...i_n} \quad \text{(summation over first indices)}$$

Tensors of rank zero are referred to as scalars. Tensors of the first rank can be represented as three-dimensional vectors and tensors of the second rank as 3×3 matrices.

A tensor A is called *isotropic* if its coordinates remain unchanged by a rotation.

In particular, the *isotropy condition* implies

$$l_{ip}l_{jq}\,a_{ij} = a_{pq} \qquad \text{for second-rank tensors } \mathsf{A} = (a_{ij}) \qquad \text{(A.7)}$$

$$l_{ip}l_{jq}l_{kr}l_{ms}\,a_{ijkm} = a_{pqrs} \quad \text{for fourth-rank tensors } \mathsf{A} = (a_{ijkm}) \qquad \text{(A.8)}$$

Problem A.2 Show that the second-rank unit tensor $\mathsf{I} = (\delta_{ij})$ is isotropic.

Lemma A.1 *A second-rank isotropic tensor* $\mathsf{A} = (a_{ij})$ *is determined by the unit tensor up to a constant a.*

$$a_{ij} = a\delta_{ij}$$

Proof Let $\mathsf{L} = (l_{mn})$ denote a rotation by the angle θ. By

$$\mathsf{L}' = (l'_{mn}) = \left(\frac{\mathrm{d}l_{mn}}{\mathrm{d}\theta}\right)$$

we denote the tensor with the components that have arisen through differentiation with respect to θ. From (A.7) it follows that

$$l_{ip}l_{jq}a_{ij} = a_{pq}$$

$$(l_{ip}l_{jq}a_{ij})' = 0$$

$$l'_{ip}l_{jq}a_{ij} + l_{ip}l'_{jq}a_{ij} = 0$$

The rotation by the angle 0 corresponds to the identity $(l_{mn}) = (\delta_{mn})$. So if we evaluate the last equation at the point $\theta = 0$, we get

$$\left[l'_{ip}a_{iq} + a_{pj}l'_{jq}\right]_{\theta=0} = 0$$

$$\left[(\mathsf{L}')^T\mathsf{A} + \mathsf{A}\mathsf{L}'\right]_{\theta=0} = 0 \qquad \text{by (A.1), (A.2)} \qquad \text{(A.9)}$$

Next, we examine rotations around the coordinate axes. For a rotation around the z-axis, we use the matrix

$$\mathsf{L}_z = \begin{bmatrix} \cos\theta & \sin\theta & 0 \\ -\sin\theta & \cos\theta & 0 \\ 0 & 0 & 1 \end{bmatrix} \quad \text{with the derivative} \quad \mathsf{L}'_z = \begin{bmatrix} -\sin\theta & \cos\theta & 0 \\ -\cos\theta & -\sin\theta & 0 \\ 0 & 0 & 0 \end{bmatrix}$$

For $\theta = 0$, we find that

$$\mathsf{L}'_z\,|_{\theta=0} = \begin{bmatrix} 0 & 1 & 0 \\ -1 & 0 & 0 \\ 0 & 0 & 0 \end{bmatrix}, \qquad (\mathsf{L}'_z\,|_{\theta=0})^T = \begin{bmatrix} 0 & -1 & 0 \\ 1 & 0 & 0 \\ 0 & 0 & 0 \end{bmatrix}$$

$$(\mathsf{L}'_z)^T \mathsf{A}\,|_{\theta=0} = \begin{bmatrix} 0 & -1 & 0 \\ 1 & 0 & 0 \\ 0 & 0 & 0 \end{bmatrix} \begin{bmatrix} a_{11} & a_{12} & a_{13} \\ a_{21} & a_{22} & a_{23} \\ a_{31} & a_{32} & a_{33} \end{bmatrix} = \begin{bmatrix} -a_{21} & -a_{22} & -a_{23} \\ a_{11} & a_{12} & a_{13} \\ 0 & 0 & 0 \end{bmatrix}$$

$$\mathsf{A}\mathsf{L}'_z\,|_{\theta=0} = \begin{bmatrix} a_{11} & a_{12} & a_{13} \\ a_{21} & a_{22} & a_{23} \\ a_{31} & a_{32} & a_{33} \end{bmatrix} \begin{bmatrix} 0 & 1 & 0 \\ -1 & 0 & 0 \\ 0 & 0 & 0 \end{bmatrix} = \begin{bmatrix} -a_{12} & a_{11} & 0 \\ -a_{22} & a_{21} & 0 \\ -a_{32} & a_{31} & 0 \end{bmatrix}$$

and therefore

$$\left[(\mathsf{L}')^T \mathsf{A} + \mathsf{A}\mathsf{L}'\right]_{\theta=0} = \begin{bmatrix} -a_{21} - a_{12} & a_{11} - a_{22} & -a_{23} \\ a_{11} - a_{22} & a_{12} + a_{21} & a_{13} \\ -a_{32} & a_{31} & 0 \end{bmatrix}$$

Because of (A.9) all components of the last matrix vanish, i.e.

$$a_{11} = a_{22}, \quad a_{23} = a_{32} = a_{31} = a_{13} = 0, \quad a_{12} = -a_{21}$$

In the same way, we examine the rotations around the x- and y-axis

$$\mathsf{L}_x = \begin{bmatrix} 1 & 0 & 0 \\ 0 & \cos\theta & -\sin\theta \\ 0 & \sin\theta & \cos\theta \end{bmatrix} \quad \text{and} \quad \mathsf{L}_y = \begin{bmatrix} \cos\theta & 0 & \sin\theta \\ 0 & 1 & 0 \\ -\sin\theta & 0 & \cos\theta \end{bmatrix}$$

In summary, we obtain

$$a_{11} = a_{22} = a_{33}, \quad a_{12} = a_{21} = a_{23} = a_{32} = a_{31} = a_{13} = 0$$

and thus the assertion. $\qquad\qquad\square$

Problem A.3 Prove that the fourth-rank tensor $\mathsf{A} = (A_{ijkm})$ with

$$A_{ijkm} = \alpha\delta_{ij}\delta_{km} + \mu\delta_{ik}\delta_{jm} + \gamma\delta_{im}\delta_{jk} \qquad (\alpha, \mu, \gamma \text{ constant})$$

is isotropic.

A.3 The Frenet Formulas

A *regular curve* $\hat{\mathbf{x}} : (a, b) \to \mathbb{R}^3$ is a twice continuously differentiable curve with

$$\hat{\mathbf{x}}'(t) \neq 0 \qquad \text{for all } t \in (a, b)$$

The vector $\hat{\mathbf{x}}'$ is called the *tangent vector*.

The *arc length*, measured from the parameter value t_0 to t, is defined by

$$s(t) = \int_{t_0}^{t} \|\hat{\mathbf{x}}'(\eta)\| \, d\eta \tag{A.10}$$

If we parameterise by the arc length, we denote the curve by $\mathbf{x}(s)$ and the tangent vector by $\mathbf{t}(s) = \mathbf{x}'(s)$.

Problem A.4 Prove that

$$\|\mathbf{t}(s)\| = 1 \tag{A.11}$$

In the following, we consider the curve in the parameterisation of its arc length.

The *normal vector* $\mathbf{n}(s)$ is the unit vector in the direction of $\mathbf{t}'(s)$. Thus, it is determined by

$$\mathbf{t}' = k(s)\,\mathbf{n} \tag{A.12}$$

The value $k(s) = \|\mathbf{x}''(s)\|$ is called the *curvature* at s.
The *binormal vector* $\mathbf{b}(s)$ is the vector product

$$\mathbf{b} = \mathbf{t} \times \mathbf{n} \tag{A.13}$$

The vectors $\mathbf{t}(s), \mathbf{n}(s), \mathbf{b}(s)$ are called the *Frenet trihedron* and form an orthonormal system at each point of the curve (see Fig. A.1).

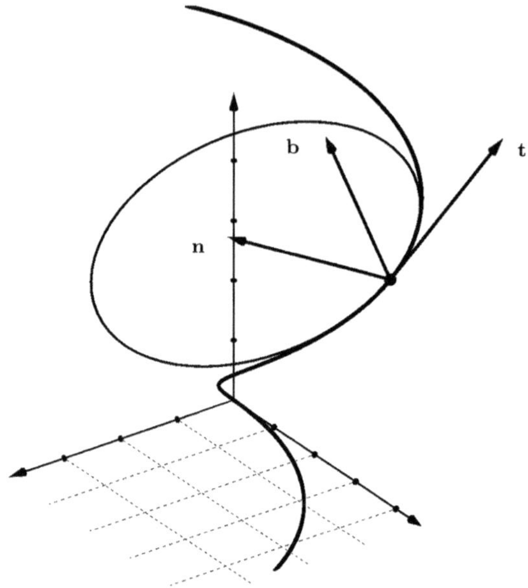

Fig. A.1 A Frenet trihedron consisting of **t**, **n** and **b**. The plane of **t** and **n** is indicated by the circle, whose curvature is $k(s)$. The lifting of the curve from the plane is described by the torsion $\tau(s)$.

At each point of the curve, the tangent and normal vectors span a plane in which the curvature $k(s)$ describes the deviation of the curve from its tangent.

Problem A.5 Prove that $\mathbf{b}'$ is collinear to **n**.

It follows that

$$\mathbf{b}'(s) = \tau(s)\,\mathbf{n}(s) \tag{A.14}$$

The parameter $\tau(s)$ defined by (A.14) is referred to as *torsion*.

Since **t**, **n** and **b** form an orthogonal system, we obtain from (A.13)

$$\mathbf{n} = \mathbf{b} \times \mathbf{t}, \qquad \mathbf{t} = \mathbf{n} \times \mathbf{b}$$

By differentiating the first equation, we get

$$\mathbf{n}' = \mathbf{b}' \times \mathbf{t} + \mathbf{b} \times \mathbf{t}' = \tau(s)\,\mathbf{n} \times \mathbf{t} + k(s)\,\mathbf{b} \times \mathbf{n} = -\tau(s)\,\mathbf{b} - k(s)\,\mathbf{t}$$

In summary, we have derived the *Frenet formulas*

$$\mathbf{t}'(s) = k(s)\,\mathbf{n}(s) \tag{A.15}$$

$$\mathbf{b}'(s) = \tau(s)\,\mathbf{n}(s) \tag{A.16}$$

$$\mathbf{n}'(s) = -\tau(s)\,\mathbf{b} - k(s)\,\mathbf{t} \tag{A.17}$$

References and Further Reading

1. Do Carmo M.: *Differential geometry of curves and surfaces*. Dover Publications, New York (2017)

A.4 Complex Numbers and Functions

Here we provide an overview of frequently used formulas for complex functions. The proofs can be found in standard works on function theory, for example in Markushevich [1]. For a presentation that focuses on fluid mechanics, we refer to Milne-Thomson [2].

According to Euler's theorem, complex numbers $z = x + \mathrm{i}y$ can be represented in the form

$$z = r\mathrm{e}^{\mathrm{i}(\phi + 2k\pi)} = r[\cos(\phi + 2k\pi) + \mathrm{i}\sin(\phi + 2k\pi)], \qquad -\pi < \phi \le \pi$$

Here we have

$$r = |z| = \sqrt{x^2 + y^2}$$

For $z \ne 0$ we obtain the argument ϕ as follows.

$$\phi = \begin{cases} \arctan \frac{y}{x} & \text{for } x > 0 \\[2ex] \arctan \frac{y}{x} + \pi & \text{for } x < 0,\ y \ge 0 \\[2ex] \arctan \frac{y}{x} - \pi & \text{for } x < 0,\ y < 0 \\[2ex] +\frac{\pi}{2} & \text{for } x = 0,\ y > 0 \\[2ex] -\frac{\pi}{2} & \text{for } x = 0,\ y < 0 \end{cases} \tag{A.18}$$

For (A.18) we write in short form

$$\phi = \arctan2\,[y, x] \tag{A.19}$$

Here it holds

$$\arctan2\,[ky, kx] = \arctan2\,[y, x] \qquad \text{for } k > 0 \tag{A.20}$$

Problem A.6 Show that dividing $z \in \mathbb{C}$ by i geometrically means that z is rotated by $90°$.

A function $f(z)$ is called *analytic* in $z_0 \in \mathbb{C}$, if it can be developed into a convergent power series with complex coefficients a_n

$$f(z) = \sum_{n=0}^{\infty} a_n (z - z_0)^n \qquad \text{in a neighbourhood of } z_0$$

The function $f(z)$ is called *analytic in the domain* $\Omega \subset \mathbb{C}$, if it is analytic at every point of the domain.

We call

$$\bar{f}(z) := \overline{f(\bar{z})}$$

the *conjugate function* to $f(z)$.

Problem A.7 Find the conjugate function to

(a) $f(z) = 3z^2 - 5iz^3$
(b) $f(z) = \ln z$

A *rectifiable curve* is a curve with finite arc length.

For the integration $\int_L f(z)\,\mathrm{d}z$ we assume that $f(z)$ is continuous in the integration domain and the curve $L \subset \mathbb{C}$ is rectifiable. If we parameterize L as

$$z = \lambda(t) = x(t) + iy(t), \qquad a \leq t \leq b,$$

where $\lambda'(t) = x'(t) + iy'(t)$ is continuous on $[a, b]$ and does not vanish anywhere, we get

$$\int_L f(z)\,\mathrm{d}z = \int_a^b f(\lambda(t))\,\lambda'(t)\,\mathrm{d}t$$

Problem A.8 Let C be the boundary of a circle with origin zero. Prove the formula

$$\oint_C \frac{dz}{z^k} = \begin{cases} 2\pi i & \text{if } k = 1 \\ 0 & \text{if } k \neq 1 \end{cases} \tag{A.21}$$

Theorem A.1 *The following statements are equivalent:*

(1) The function $f(z)$ is analytic in a circle $K \subset \mathbb{C}$.

(2) In the representation $f(z) = u(x, y) + \mathrm{i}v(x, y)$, the functions u, v are twice differentiable in K and satisfy the Cauchy-Riemann differential equations

$$\frac{\partial u}{\partial x} = \frac{\partial v}{\partial y}, \qquad \frac{\partial u}{\partial y} = -\frac{\partial v}{\partial x}$$

(3) The function $f(z)$ is continuous in K. For every simple closed curve $C \subset K$ it holds that

$$\oint_C f(z)\, dz = 0$$

Theorem A.2 (Laurent) *A function f that is analytic in the annular domain*

$$\Omega = \{z \in \mathbb{C} \mid r < |z - z_0| < R\}$$

can be represented as a convergent Laurent series

$$f(z) = \sum_{-\infty}^{\infty} a_n (z - z_0)^n$$

For $r = 0$, $R = \infty$ we obtain the special case of the complex plane from which a point has been removed.

The following concept introduces suitable coordinates for integrals around airfoil profiles.

A continuous and bijective mapping of two domains of the complex plane is called *conformal* at a point z if the angle of intersection of two curves through z is preserved in the mapping.

Problem A.9 Prove that an analytic function $f(z)$ is conformal at z_0 if $f'(z_0) \neq 0$ holds.

References and Further Reading

1. Markushevich A.I.: *The Theory of Analytic Functions. A Brief Course.* Mir Publishers, Moscow (1983)
2. Milne-Thomson L.M.: *Theoretical Hydrodynamics.* Dover Publications, New York (1968)

A.5 Non-Stationarity in Turbulent Flows

In Sect. 3.5 it was mentioned that turbulence is always unsteady. We demonstrate this property by modeling the turbulence in a highly simplified form using a system of potential vortices. The cause of the instationarity turns out to be the mutual displacement of the vortex centers. Depending on the signs of the circulation in the involved vortices, they are induced to different movements, which we illustrate with two simple examples. For the modeling, we use complex potentials, the basics of which are explained in the Sects. 4.2 and 4.4.

Co-directed Potential Vortices

First, we consider a flow consisting of two (co-directed) potential vortices with circulations Γ_1, $\Gamma_2 > 0$, whose centres are located at z_1 and z_2 at time $t = 0$ (Fig. A.2). According to (4.16), its potential is given by

$$w(z) = \frac{i\Gamma_1}{2\pi} \ln(z - z_1) + \frac{i\Gamma_2}{2\pi} \ln(z - z_2) \tag{A.22}$$

The vortex at z_1 causes a rotation of all points $z \neq z_1$, including z_2. Similarly, the vortex at z_2 acts on the centre z_1.

Fig. A.2 Co-directed potential vortices of different strengths generate a rotation around the centroid. The distance between the vortex centers remains constant.

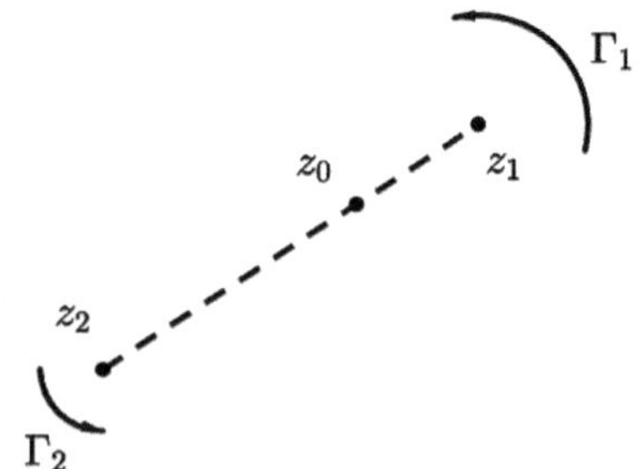

Using the approach

$$z_1 = z_1(t), \quad z_2 = z_2(t) \qquad t \geq 0,$$

we examine the trajectories of the vortex centres, whose velocities are denoted by $\dot{z}_1$ and $\dot{z}_2$. According to Sect. 4.2, we obtain the complex velocity at z_2 from the first term in (A.22)

$$\mathcal{U}(z_2) = u_2 - iv_2 = \frac{\mathrm{d}}{\mathrm{d}z} \frac{i\Gamma_1}{2\pi} \ln(z - z_1)\Big|_{z=z_2} = \frac{i\Gamma_1}{2\pi(z_2 - z_1)}$$

and thus

$$\dot{z}_2 = u_2 + iv_2 = \overline{u_2 - iv_2} = \frac{-i\Gamma_1}{2\pi(\bar{z}_2 - \bar{z}_1)} \tag{A.23}$$

Similarly, we have

$$\dot{z}_1 = \frac{-i\Gamma_2}{2\pi(\bar{z}_1 - \bar{z}_2)} \tag{A.24}$$

In analogy to the center of mass of two bodies, we define the *centroid* as the weighted mean center of the vortex centres

$$z_0 = \frac{\Gamma_1 z_1 + \Gamma_2 z_2}{\Gamma_1 + \Gamma_2} \tag{A.25}$$

Let us examine its motion.

Problem A.10 Prove the following statements

(a)

$$\Gamma_1 \dot{z}_1 + \Gamma_2 \dot{z}_2 = 0 \tag{A.26}$$

(b)   The position of the centroid remains unchanged, i.e. it is

$$\dot{z}_0 = 0$$

(c)   The distance between the vortex centers remains unchanged, i.e. it is

$$\frac{\mathrm{d}}{\mathrm{d}t} |z_1 - z_2| = 0$$

(d)   The system rotates around the centroid z_0 with the angular velocity

$$\omega = \frac{\Gamma_1 + \Gamma_2}{|z_1 - z_2|^2}$$

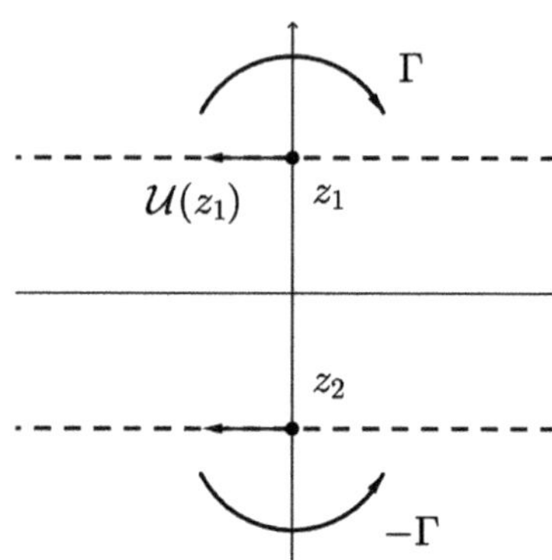

Fig. A.3 Oppositely directed potential vortices of equal strength lead to a shift of the vortex centers parallel to the axis of symmetry.

In particular, this flow is *unsteady*. The same applies to turbulent areas containing a multitude of vortices.

Oppositely Directed Potential Vortices

In the example of two opposite potential vortices, we limit ourselves to the special case that they have the same strength. We choose the coordinate system such that the vortex centres are located at positions $z_1 = a\mathrm{i}$ (for $\Gamma_1 = \Gamma$) and $z_2 = -a\mathrm{i}$ (for $\Gamma_2 = -\Gamma$) at time $t = 0$ (Fig. A.3). According to (4.16), their potential is given by

$$w(z) = \frac{\mathrm{i}\Gamma}{2\pi} \ln(z - a\mathrm{i}) - \frac{\mathrm{i}\Gamma}{2\pi} \ln(z + a\mathrm{i}) \qquad \text{for } t = 0 \tag{A.27}$$

Each vortex exerts the same force on the opposite vortex centre, thereby inducing a velocity. Owing to symmetry, both vortex centres move with the same velocity

$$\mathcal{U}(z_1) = \frac{\mathrm{d}}{\mathrm{d}z}\left[-\frac{\mathrm{i}\Gamma}{2\pi} \ln(z + a\mathrm{i}) \right]\Bigg|_{z=z_1} = -\frac{\Gamma}{4\pi a}$$

in the x-direction and we obtain an *unsteady flow* of two parallel vortices.

Just like (A.23), (A.24), the motion of the vortex centres in (A.27) can be formulated as a nonlinear system of ordinary differential equations. The more potential vortices interact, the larger the dimension of the system becomes, and it is to be expected that the solution will exhibit chaotic behavior.

A.6   Flow Separation on the Cylinder

In Fig. 5.5 we illustrated the early flow separation on a circular cylinder. Here we derive a corresponding complex potential with two stationary potential vortices, whose centres are located behind the cylinder. Since oppositely directed potential vortices of equal strength move linearly and uniformly according to the previous section, a stationary state can be achieved by inflow from the opposite direction.

The starting point is formula (4.19) for the potential of the flow around the cylinder $|z| = a$. For horizontal inflow ($\alpha = 0$) it takes the form

$$w_1(z) = u_\infty \left[z + \frac{a^2}{z} \right] \qquad \text{(vortex-free cylinder potential)}$$

Now we consider the flow of two potential vortices with oppositely directed circulations of equal strenght. Their centers are symmetric to the x-axis and outside the cylinder. In analogy with (A.27) we describe the potential by

$$w_p(z) = \frac{i\Gamma}{2\pi} \ln(z - z_1) - \frac{i\Gamma}{2\pi} \ln(z - \bar{z}_1) \qquad (A.28)$$

Since the boundary $|z| = a$ does not form a streamline for w_p, the potential w_p and thus also $w_1 + w_p$ are unsuitable for flow around the cylinder. To correct this, we modify it using the Milne-Thomson theorem (Theorem 4.4).

$$w_2(z) = w_p(z) + \bar{w}_p\left(\frac{a^2}{z}\right)$$

Then the following sum is also a cylinder potential.

$$\begin{aligned}
w(z) &= w_1(z) + w_p(z) + \bar{w}_p\left(\frac{a^2}{z}\right) \\
&= u_\infty \left[z + \frac{a^2}{z} \right] + \frac{i\Gamma}{2\pi}[\ln(z - z_1) - \ln(z - \bar{z}_1)] \\
&\quad + \underbrace{\frac{i\Gamma}{2\pi}\left[\ln\left(\frac{a^2}{z} - \bar{z}_1\right) - \ln\left(\frac{a^2}{z} - z_1\right) \right]}_{\text{bound vortex}} \qquad (A.29)
\end{aligned}$$

For $|z_1| > a$ it holds that $|a^2/z_1| = |a^2/\bar{z}_1| < a$. As a result, the centers of the last vortex term are inside the cylinder, i.e. it is a bound vortex.

Since a potential vortex does not change its own centre, the velocity in z_1 is determined only by the vortex-free cylinder potential, the potential vortex at $\bar{z}_1$ and the bound vortex. They represent the *part of the potential that is relevant for the movement of z_1*:

$$\hat{w}(z) = u_\infty \left[z + \frac{a^2}{z} \right] + \frac{i\Gamma}{2\pi}\left[\ln\left(\frac{a^2}{z} - \bar{z}_1\right) - \ln\left(\frac{a^2}{z} - z_1\right) - \ln(z - \bar{z}_1) \right]$$

Thus, the vortex centre z_1 moves with the velocity

$$\mathcal{U}(z_1) = \frac{\mathrm{d}}{\mathrm{d}z}\hat{w}(z)\,|_{z=z_1}$$

$$= u_\infty\left[1 - \frac{a^2}{z_1^2}\right] + \frac{\mathrm{i}\Gamma}{2\pi}\left[\frac{z_1}{a^2 - z_1^2} - \frac{\bar{z}_1}{a^2 - z_1\bar{z}_1} - \frac{1}{z_1 - \bar{z}_1}\right]$$

$$= u_\infty\left[1 - \frac{a^2}{z_1^2}\right] + \frac{\mathrm{i}\Gamma}{2\pi}\frac{a^2(z_1 - \bar{z}_1)^2 + (z_1^2 - a^2)(z_1\bar{z}_1 - a^2)}{(a^2 - z_1^2)(a^2 - z_1\bar{z}_1)(z_1 - \bar{z}_1)}$$

The condition for a stationary flow is

$$\mathcal{U}(z_1) = 0$$

$$u_\infty\frac{z_1^2 - a^2}{z_1^2} = \frac{\mathrm{i}\Gamma}{2\pi}\frac{a^2(z_1 - \bar{z}_1)^2 + (z_1^2 - a^2)(z_1\bar{z}_1 - a^2)}{(a^2 - z_1^2)(a^2 - z_1\bar{z}_1)(z_1 - \bar{z}_1)}$$

If we form the complex conjugate on both sides, we get

$$u_\infty\frac{\bar{z}_1^2 - a^2}{\bar{z}_1^2} = -\frac{\mathrm{i}\Gamma}{2\pi}\frac{a^2(\bar{z}_1 - z_1)^2 + (\bar{z}_1^2 - a^2)(z_1\bar{z}_1 - a^2)}{(a^2 - \bar{z}_1^2)(a^2 - z_1\bar{z}_1)(\bar{z}_1 - z_1)}$$

If we divide the second to last equation by the last one, we get after rearranging

$$\frac{(z_1^2 - a^2)^2\,\bar{z}_1^2}{(\bar{z}_1^2 - a^2)^2\,z_1^2} = \frac{(z_1^2 - a^2)(z_1\bar{z}_1 - a^2) + a^2(z_1 - \bar{z}_1)^2}{(\bar{z}_1^2 - a^2)(z_1\bar{z}_1 - a^2) + a^2(z_1 - \bar{z}_1)^2}$$

$$0 = (z_1\bar{z}_1 - a^2)^2 + z_1\bar{z}_1(z_1 - \bar{z}_1)^2$$

Using polar coordinates $z_1 = r\mathrm{e}^{\mathrm{i}\theta}$, $0 < \theta < \pi/2$, we find that

$$(r^2 - a^2)^2 = 4r^4\sin^2\theta$$

$$r - \frac{a^2}{r} = 2r\sin\theta$$

$$\left|z_1 - \frac{a^2}{z_1}\right| = |z_1 - \bar{z}_1| \qquad \text{see Fig. A.4} \qquad (A.30)$$

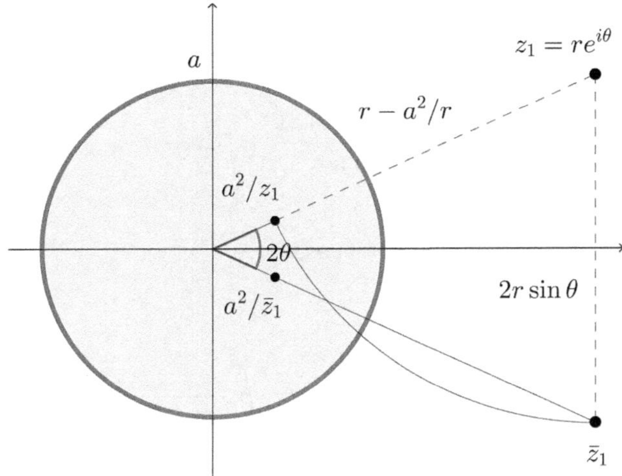

Fig. A.4 The potential (A.29) has four vortices. In the stationary case, the vortex centre z_1 has the same distance from a^2/z_1 as from $\bar{z}_1$.

Consequently, our cylinder potential is given by (A.29), where the position of the vortex center z_1 is determined by its distance (A.30) to the centers a^2/z_1 and $\bar{z}_1$.

A.7   Glauert Integrals

Theorem A.3 *It holds*

$$\int_0^\pi \frac{\cos nx}{\cos x - \cos\theta}\, \mathrm{d}x = \frac{\pi \sin n\theta}{\sin\theta} \tag{A.31}$$

Proof Since the integrand is even, we have

$$\int_0^\pi \frac{\cos nx}{\cos x - \cos\theta}\, \mathrm{d}x = \frac{1}{2}\int_{-\pi}^\pi \frac{\cos nx}{\cos x - \cos\theta}\, \mathrm{d}x$$

From

$$I = \int_{-\pi}^\pi \frac{\cos nx}{\cos x - \cos\theta}\, \mathrm{d}x, \quad J = \int_{-\pi}^\pi \frac{\sin nx}{\cos x - \cos\theta}\, \mathrm{d}x$$

we get

$$\int_0^\pi \frac{\cos nx}{\cos x - \cos\theta}\, \mathrm{d}x = \frac{1}{2}\operatorname{Re}(I + \mathrm{i}J)$$

For $z = e^{ix} = \cos x + i \sin x$ we have

$$dz = -\sin x \, dx + i \cos x \, dx = (-\sin x + i \cos x) \, dx$$

$$\frac{dz}{iz} = \frac{(-\sin x + i \cos x) \, dx}{i \cos x - \sin x} = dx$$

$$\frac{z + z^{-1}}{2} = \frac{e^{ix} + e^{-ix}}{2} = \cos x$$

and therefore

$$I + iJ = \int_{-\pi}^{\pi} \frac{\cos nx + i \sin nx}{\cos x - \cos \theta} \, dx = \oint \frac{z^n}{\frac{z+z^{-1}}{2} - \cos \theta} \frac{dz}{iz}$$

$$= 2i \oint \frac{z^n}{2i\left(\frac{z+z^{-1}}{2} - \cos \theta\right)} \frac{dz}{iz} = -2i \oint \frac{z^n}{2z\left(\frac{z+z^{-1}}{2} - \cos \theta\right)} \, dz$$

$$= -2i \oint \frac{z^n}{z^2 + 1 - 2z \cos \theta} \, dz$$

Now we look for the poles of $f(z) = \frac{z^n}{z^2+1-2z \cos \theta}$.

$$0 = z^2 - 2z \cos \theta + 1$$

$$z = \cos \theta \pm \sqrt{\cos^2 \theta - 1} = \cos \theta \pm i \sin \theta$$

Thus we get the poles at

$$z_1 = \cos \theta + i \sin \theta = e^{i\theta}, \qquad z_2 = \cos \theta - i \sin \theta = e^{-i\theta}$$

Therefore,

$$f(z) = \frac{z^n}{(z - z_1)(z - z_2)}$$

Now we calculate the residues. Recall that the poles lie on the boundary $|z| = 1$.

$$\operatorname{Res}(f(z), z_1) = \lim_{z \to z_1} (z - z_1) f(z) = \frac{z_1^n}{z_1 - z_2} = \frac{e^{in\theta}}{e^{i\theta} - e^{-i\theta}} = \frac{e^{in\theta}}{2i \sin \theta}$$

$$\operatorname{Res}(f(z), z_2) = \lim_{z \to z_2} (z - z_2) f(z) = \frac{z_2^n}{z_2 - z_1} = \frac{e^{-in\theta}}{e^{-i\theta} - e^{i\theta}} = -\frac{e^{-in\theta}}{2i \sin \theta}$$

$$\sum \operatorname{Res} = \frac{e^{in\theta}}{2i \sin \theta} - \frac{e^{-in\theta}}{2i \sin \theta} = \frac{e^{in\theta} - e^{-in\theta}}{2i \sin \theta} = \frac{2i \sin n\theta}{2i \sin \theta} = \frac{\sin n\theta}{\sin \theta}$$

Applying the residue theorem for poles on the circle boundary (see [1]), we evaluate the circulation integral.

$$\oint \frac{z^n}{z^2 + 1 - 2z\cos\theta}\, dz = \pi i \sum \text{Res} = \pi i \frac{\sin n\theta}{\sin\theta}$$

Then it follows that

$$\int_0^\pi \frac{\cos nx}{\cos x - \cos\theta}\, dx = \frac{1}{2}\text{Re}\left(-2i \oint \frac{z^n}{z^2 + 1 - 2z\cos\theta}\, dz\right)$$

$$= \text{Re}\left(\pi \frac{\sin n\theta}{\sin\theta}\right) = \pi \frac{\sin n\theta}{\sin\theta}$$

$\square$

References and Further Reading

1. Legua M., Sanchez-Ruiz L.M.: *Cauchy Principal Value Contour Integral with Applications.* Entropy (2017), 19, 215; https://doi.org/10.3390/e19050215.

A.8   Cylindrical Coordinates

First, we look for formulas for converting between the Cartesian coordinate system (x_1, x_2, x_3) with unit vectors $\mathbf{i}_1, \mathbf{i}_2, \mathbf{i}_3$ and an arbitrary orthogonal coordinate system (ξ_1, ξ_2, ξ_3) with unit vectors $\mathbf{e}_1, \mathbf{e}_2, \mathbf{e}_3$. This allows us to rewrite the Navier-Stokes equations in cylindrical coordinates.

For the differential $\mathbf{dr} = \sum_k \mathbf{i}_k\, dx_k$ it follows from $dx_k = \sum_l \frac{\partial x_k}{\partial \xi_l}\, d\xi_l$ that

$$\mathbf{dr} = \sum_{k,l} \left(\mathbf{i}_k \frac{\partial x_k}{\partial \xi_l}\right) d\xi_l$$

By introducing the quantities

$$h_l = \left\|\sum_k \mathbf{i}_k \frac{\partial x_k}{\partial \xi_l}\right\| = \sqrt{\left(\frac{\partial x_1}{\partial \xi_l}\right)^2 + \left(\frac{\partial x_2}{\partial \xi_l}\right)^2 + \left(\frac{\partial x_3}{\partial \xi_l}\right)^2} \qquad (A.32)$$

we obtain

$$\mathbf{dr} = \sum_l h_l \mathbf{e}_l\, d\xi_l \qquad (A.33)$$

with the new unit vectors

$$\mathbf{e}_l = \frac{1}{h_l} \sum_k \mathbf{i}_k \frac{\partial x_k}{\partial \xi_l} \tag{A.34}$$

The derivation of the transformation rules is given in several stages, which we formulate as problems.

Problem A.11 Prove the gradient representation

$$\nabla = \sum_n \frac{1}{h_n} \mathbf{e}_n \frac{\partial}{\partial \xi_n} \tag{A.35}$$

as well as

$$\nabla \xi_k = \frac{1}{h_k} \mathbf{e}_k \tag{A.36}$$

Problem A.12 Prove

$$\nabla \times \mathbf{e}_k = h_k \mathbf{e}_k \times \left(\nabla \frac{1}{h_k} \right) \tag{A.37}$$

Problem A.13 Prove

$$\begin{aligned}
\nabla \times \mathbf{e}_1 &= -\frac{1}{h_1 h_2} \frac{\partial h_1}{\partial \xi_2} \mathbf{e}_3 + \frac{1}{h_1 h_3} \frac{\partial h_1}{\partial \xi_3} \mathbf{e}_2 \\
\nabla \times \mathbf{e}_2 &= \frac{1}{h_1 h_2} \frac{\partial h_2}{\partial \xi_1} \mathbf{e}_3 - \frac{1}{h_3 h_2} \frac{\partial h_2}{\partial \xi_3} \mathbf{e}_1 \\
\nabla \times \mathbf{e}_3 &= -\frac{1}{h_1 h_3} \frac{\partial h_3}{\partial \xi_1} \mathbf{e}_2 + \frac{1}{h_3 h_2} \frac{\partial h_3}{\partial \xi_2} \mathbf{e}_1
\end{aligned} \tag{A.38}$$

Problem A.14 Prove

$$\nabla \mathbf{e}_1 = \frac{1}{h_1 h_2 h_3} \frac{\partial (h_2 h_3)}{\partial \xi_1}, \quad \nabla \mathbf{e}_2 = \frac{1}{h_1 h_2 h_3} \frac{\partial (h_1 h_3)}{\partial \xi_2}, \quad \nabla \mathbf{e}_3 = \frac{1}{h_1 h_2 h_3} \frac{\partial (h_1 h_2)}{\partial \xi_3} \tag{A.39}$$

Problem A.15 Consider the function $\mathbf{q} = \sum_k q_k \mathbf{e}_k$. Prove

$$\operatorname{div} \mathbf{q} = \frac{1}{h_1 h_2 h_3} \left[\frac{\partial (q_1 h_2 h_3)}{\partial \xi_1} + \frac{\partial (q_2 h_1 h_3)}{\partial \xi_2} + \frac{\partial (q_3 h_1 h_2)}{\partial \xi_3} \right] \tag{A.40}$$

Problem A.16 Prove

$$\Delta\phi = \frac{1}{h_1 h_2 h_3}\left[\frac{\partial}{\partial\xi_1}\left(\frac{h_2 h_3}{h_1}\frac{\partial\phi}{\partial\xi_1}\right) + \frac{\partial}{\partial\xi_2}\left(\frac{h_1 h_3}{h_2}\frac{\partial\phi}{\partial\xi_2}\right) + \frac{\partial}{\partial\xi_3}\left(\frac{h_1 h_2}{h_3}\frac{\partial\phi}{\partial\xi_3}\right)\right] \qquad \text{(A.41)}$$

Now we can transform the Navier-Stokes equations. When transitioning from Cartesian to cylindrical coordinates, we use the notation

$$(x_1, x_2, x_3) = (x, y, z); \qquad (\xi_1, \xi_2, \xi_3) = (r, \theta, z)$$

and the formulas

$$x = r\cos\theta, \qquad y = r\sin\theta$$

We obtain from (A.32)

$$\begin{aligned}
h_1 &= \sqrt{\cos^2\theta + \sin^2\theta} = 1 \\
h_2 &= \sqrt{r^2\sin^2\theta + r^2\cos^2\theta} = r \\
h_3 &= 1
\end{aligned} \qquad \text{(A.42)}$$

From (A.34) we find the new unit vectors (see Fig. A.5)

$$\mathbf{e}_r = \mathbf{i}_x\cos\theta + \mathbf{i}_y\sin\theta, \qquad \mathbf{e}_\theta = -\mathbf{i}_x\sin\theta + \mathbf{i}_y\cos\theta$$

The third vector remains unchanged, i.e. $\mathbf{e}_z = \mathbf{i}_z$. For the derivatives we obtain

$$\frac{\partial\mathbf{e}_r}{\partial r} = \qquad\qquad \frac{\partial\mathbf{e}_\theta}{\partial r} = \frac{\partial\mathbf{e}_z}{\partial r} = 0 \qquad \text{(A.43)}$$

$$\frac{\partial\mathbf{e}_r}{\partial z} = \qquad\qquad \frac{\partial\mathbf{e}_\theta}{\partial z} = \frac{\partial\mathbf{e}_z}{\partial z} = 0 \qquad \text{(A.44)}$$

$$\frac{\partial\mathbf{e}_r}{\partial\theta} = \qquad \mathbf{e}_\theta, \qquad \frac{\partial\mathbf{e}_\theta}{\partial\theta} = -\mathbf{e}_r, \qquad \frac{\partial\mathbf{e}_z}{\partial\theta} = 0 \qquad \text{(A.45)}$$

Using (A.35) and (A.42) we find the pressure gradient

$$\nabla p = \frac{\partial p}{\partial r}\mathbf{e}_r + \frac{1}{r}\frac{\partial p}{\partial\theta}\mathbf{e}_\theta + \frac{\partial p}{\partial z}\mathbf{e}_z \qquad \text{(A.46)}$$

In cylindrical coordinates, we denote the velocity vector by

$$\mathbf{u} = u_r\mathbf{e}_r + u_\theta\mathbf{e}_\theta + u_z\mathbf{e}_z$$

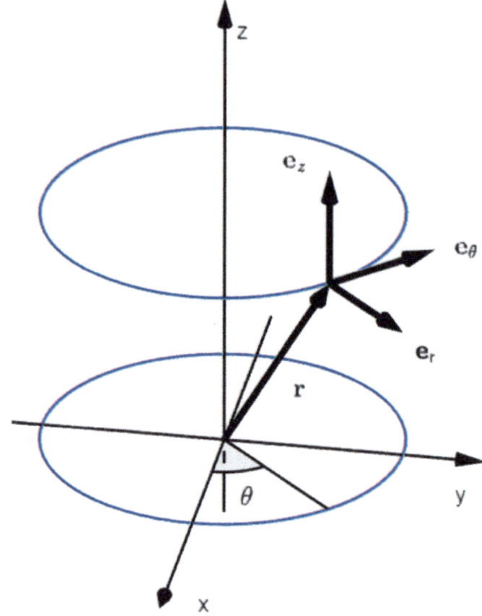

Fig. A.5 In the Cartesian coordinate system, the cylindrical unit vectors $\mathbf{e}_r$, $\mathbf{e}_\theta$ depend on the considered spatial point.

Because $\mathbf{u} = d\mathbf{r}/dt$ it follows from (A.33) that

$$\mathbf{u} = \frac{dr}{dt}\mathbf{e}_r + r\frac{d\theta}{dt}\mathbf{e}_\theta + \frac{dz}{dt}\mathbf{e}_z$$

and thus

$$u_r = \frac{dr}{dt}, \qquad u_\theta = r\frac{d\theta}{dt}, \qquad u_z = \frac{dz}{dt} \tag{A.47}$$

From (A.40) and (A.42) we obtain the divergence of the velocity field

$$\operatorname{div}\mathbf{u} = \frac{\partial u_r}{\partial r} + \frac{1}{r}\frac{\partial u_\theta}{\partial \theta} + \frac{\partial u_z}{\partial z} + \frac{u_r}{r} \tag{A.48}$$

Problem A.17 Prove for the material derivative that

$$\begin{aligned}
\frac{D\mathbf{u}}{Dt} &= \left[\dot{u}_r + u_r\frac{\partial u_r}{\partial r} + \frac{u_\theta}{r}\frac{\partial u_r}{\partial \theta} - \frac{u_\theta^2}{r} + u_z\frac{\partial u_r}{\partial z}\right]\mathbf{e}_r \\
&\quad + \left[\dot{u}_\theta + u_r\frac{\partial u_\theta}{\partial r} + \frac{u_\theta}{r}\frac{\partial u_\theta}{\partial \theta} + \frac{u_\theta u_r}{r} + u_z\frac{\partial u_\theta}{\partial z}\right]\mathbf{e}_\theta \\
&\quad + \left[\dot{u}_z + u_r\frac{\partial u_z}{\partial r} + \frac{u_\theta}{r}\frac{\partial u_z}{\partial \theta} + u_z\frac{\partial u_z}{\partial z}\right]\mathbf{e}_z
\end{aligned} \tag{A.49}$$

By (A.41) we get the Laplace operator

$$\Delta \phi = \frac{1}{r}\left[\frac{\partial}{\partial r}\left(r\frac{\partial \phi}{\partial r}\right) + \frac{\partial}{\partial \theta}\left(\frac{1}{r}\frac{\partial \phi}{\partial \theta}\right) + \frac{\partial}{\partial z}\left(r\frac{\partial \phi}{\partial z}\right)\right]$$

$$= \frac{\partial^2 \phi}{\partial r^2} + \frac{1}{r}\frac{\partial \phi}{\partial r} + \frac{1}{r^2}\frac{\partial^2 \phi}{\partial \theta^2} + \frac{\partial^2 \phi}{\partial z^2} \tag{A.50}$$

Problem A.18 Prove for the diffusion term that

$$\Delta \mathbf{u} = \left(\Delta u_r - \frac{u_r}{r^2} - \frac{2}{r^2}\frac{\partial u_\theta}{\partial \theta}\right)\mathbf{e}_r + \left(\Delta u_\theta + \frac{2}{r^2}\frac{\partial u_\theta}{\partial \theta} - \frac{u_\theta}{r^2}\right)\mathbf{e}_\theta + \Delta u_z\,\mathbf{e}_z \tag{A.51}$$

Using the notation

$$\frac{Du_r}{Dt} = \frac{\partial u_r}{\partial t} + u_r\frac{\partial u_r}{\partial r} + \frac{u_\theta}{r}\frac{\partial u_r}{\partial \theta} + u_z\frac{\partial u_r}{\partial z} \tag{A.52}$$

$$\frac{Du_\theta}{Dt} = \frac{\partial u_\theta}{\partial t} + u_r\frac{\partial u_\theta}{\partial r} + \frac{u_\theta}{r}\frac{\partial u_\theta}{\partial \theta} + u_z\frac{\partial u_\theta}{\partial z} \tag{A.53}$$

$$\frac{Du_z}{Dt} = \frac{\partial u_z}{\partial t} + u_r\frac{\partial u_z}{\partial r} + \frac{u_\theta}{r}\frac{\partial u_z}{\partial \theta} + u_z\frac{\partial u_z}{\partial z}, \tag{A.54}$$

we obtain from (A.46), (A.48), (A.49), (A.51) the Navier-Stokes equations in cylindrical coordinates

$$\frac{Du_r}{Dt} - \frac{u_\theta^2}{r} + \frac{1}{\varrho}\frac{\partial p}{\partial r} = v\left(\Delta u_r - \frac{u_r}{r^2} - \frac{2}{r^2}\frac{\partial u_\theta}{\partial \theta}\right) \tag{A.55}$$

$$\frac{Du_\theta}{Dt} + \frac{u_r u_\theta}{r} + \frac{1}{\varrho r}\frac{\partial p}{\partial \theta} = v\left(\Delta u_\theta + \frac{2}{r^2}\frac{\partial u_\theta}{\partial \theta} - \frac{u_\theta}{r^2}\right) \tag{A.56}$$

$$\frac{Du_z}{Dt} + \frac{1}{\varrho}\frac{\partial p}{\partial z} = v\Delta u_z \tag{A.57}$$

$$\frac{\partial u_r}{\partial r} + \frac{1}{r}\frac{\partial u_\theta}{\partial \theta} + \frac{\partial u_z}{\partial z} + \frac{u_r}{r} = 0 \tag{A.58}$$

A.9   A Derivation of the Biot-Savart Law

For a given vorticity field $\omega(\mathbf{x})$, the corresponding velocity field $\mathbf{u}(\mathbf{x}, t)$ can be determined up to the addition of a divergence-free component.

Theorem A.4 *Let $\omega(\mathbf{x})$ be a smooth vorticity field that vanishes for $\|\mathbf{x}\| > R$ according to*

$$\|\omega(\mathbf{x})\| \leq C\|\mathbf{x}\|^{-3}$$

If the corresponding velocity field **u** *is divergence-free and satisfies the condition*

$$\lim_{\|\mathbf{x}\| \to 0} \|\mathbf{u}\| = 0,$$

then the only solution is given by

$$\mathbf{u}(\mathbf{x}) = \frac{1}{4\pi} \int_{\mathbb{R}^3} \frac{(\mathbf{y} - \mathbf{x}) \times \boldsymbol{\omega}(\mathbf{y})}{\|\mathbf{x} - \mathbf{y}\|^3} \, dV_{\mathbf{y}}$$

The following proof is used in electrodynamics to derive the Biot-Savart law from Maxwell's equations.

Proof The Poisson equation $\Delta \mathbf{v} = -\boldsymbol{\omega}$ can be written out component-wise and solved with a scalar Newton potential (see [1]).

$$\Delta v_i = -\omega_i \quad \Rightarrow \quad v_i(\mathbf{x}) = \frac{1}{4\pi} \int_{\mathbb{R}^3} \frac{\omega_i(\mathbf{y})}{\|\mathbf{x} - \mathbf{y}\|} \, dV_{\mathbf{y}} \qquad (i = 1, 2, 3)$$

Here, we ensure the existence of the integrals by the growth condition on $\boldsymbol{\omega}$. We combine the potential components into a vector potential.

$$\mathbf{v}(\mathbf{x}) = \frac{1}{4\pi} \int_{\mathbb{R}^3} \frac{\boldsymbol{\omega}(\mathbf{y})}{\|\mathbf{x} - \mathbf{y}\|} \, dV_{\mathbf{y}}$$

Applying the curl, we get

$$\operatorname{rot} \mathbf{v} = \frac{1}{4\pi} \int_{\mathbb{R}^3} \operatorname{rot}_{\mathbf{x}} \frac{\boldsymbol{\omega}(\mathbf{y})}{\|\mathbf{x} - \mathbf{y}\|} \, dV_{\mathbf{y}}$$

Using (1.13) with $\alpha = \frac{1}{\|\mathbf{x} - \mathbf{y}\|}$ and $\mathbf{a} = \boldsymbol{\omega}(\mathbf{y})$, we calculate $\operatorname{rot}_{\mathbf{x}} \frac{\boldsymbol{\omega}(\mathbf{y})}{\|\mathbf{x} - \mathbf{y}\|}$. Here we write $r = \|\mathbf{y} - \mathbf{x}\|$ in an intermediate step.

$$
\begin{aligned}
\operatorname{rot}_{\mathbf{x}} \frac{\boldsymbol{\omega}(\mathbf{y})}{\|\mathbf{x} - \mathbf{y}\|} &= \left(\operatorname{grad}_{\mathbf{x}} \frac{1}{\|\mathbf{x} - \mathbf{y}\|} \right) \times \boldsymbol{\omega}(\mathbf{y}) + \frac{1}{\|\mathbf{x} - \mathbf{y}\|} \operatorname{rot}_{\mathbf{x}} \boldsymbol{\omega}(\mathbf{y}) \\
&= \left(\operatorname{grad}_{\mathbf{x}} \frac{1}{\|\mathbf{x} - \mathbf{y}\|} \right) \times \boldsymbol{\omega}(\mathbf{y}) \quad \text{because } \boldsymbol{\omega}(\mathbf{y}) \text{ does not depend on } \mathbf{x} \\
&= \frac{\mathbf{y} - \mathbf{x}}{\|\mathbf{x} - \mathbf{y}\|^3} \times \boldsymbol{\omega}(\mathbf{y}) \quad \text{using (1.30) for } f(r) = 1/r
\end{aligned}
$$

If we insert this expression into the last integral, we obtain

$$\operatorname{rot} \mathbf{v} = \frac{1}{4\pi} \int_{\mathbb{R}^3} \frac{(\mathbf{y} - \mathbf{x}) \times \boldsymbol{\omega}(\mathbf{y})}{\|\mathbf{x} - \mathbf{y}\|^3} \, dV_{\mathbf{y}}$$

We want to check whether $\mathrm{rot}\,\mathbf{v}$ is indeed a velocity field for $\boldsymbol{\omega}$. For this purpose we need to show $\mathrm{rot}\,(\mathrm{rot}\,\mathbf{v}) = \boldsymbol{\omega}$.

$$\mathrm{rot}\,(\mathrm{rot}\,\mathbf{v}) = \mathrm{grad}\,(\mathrm{div}\,\mathbf{v}) - \Delta\mathbf{v} \quad \text{by (1.14)}$$
$$= \mathrm{grad}\,(\mathrm{div}\,\mathbf{v}) + \boldsymbol{\omega}, \tag{A.59}$$

where we have used the above Poisson equation. Now we consider the integral

$$\mathrm{div}\,\mathbf{v} = \frac{1}{4\pi} \int_{\mathbb{R}^3} \mathrm{div}_{\mathbf{x}} \frac{\boldsymbol{\omega}(\mathbf{y})}{\|\mathbf{x} - \mathbf{y}\|}\, dV_{\mathbf{y}} \tag{A.60}$$

Using (1.15) with $\alpha = \frac{1}{\|\mathbf{x}-\mathbf{y}\|}$ and $\mathbf{a} = \boldsymbol{\omega}(\mathbf{y})$, we calculate

$$\mathrm{div}_{\mathbf{x}} \frac{\boldsymbol{\omega}(\mathbf{y})}{\|\mathbf{x} - \mathbf{y}\|} = \frac{1}{\|\mathbf{x} - \mathbf{y}\|} \mathrm{div}_{\mathbf{x}}\,\boldsymbol{\omega}(\mathbf{y}) + \boldsymbol{\omega}(\mathbf{y}) \cdot \mathrm{grad}_{\mathbf{x}} \frac{1}{\|\mathbf{x} - \mathbf{y}\|}$$
$$= \boldsymbol{\omega}(\mathbf{y}) \cdot \mathrm{grad}_{\mathbf{x}} \frac{1}{\|\mathbf{x} - \mathbf{y}\|} \quad \text{because } \boldsymbol{\omega}(\mathbf{y}) \text{ does not depend on } \mathbf{x}$$
$$= -\boldsymbol{\omega}(\mathbf{y}) \cdot \mathrm{grad}_{\mathbf{y}} \frac{1}{\|\mathbf{x} - \mathbf{y}\|} \quad \text{by (1.28), (1.29)}$$

Applying (1.15), we find that

$$\mathrm{div}_{\mathbf{y}} \frac{\boldsymbol{\omega}(\mathbf{y})}{\|\mathbf{x} - \mathbf{y}\|} = \frac{1}{\|\mathbf{x} - \mathbf{y}\|} \mathrm{div}_{\mathbf{y}}\,\boldsymbol{\omega}(\mathbf{y}) + \boldsymbol{\omega}(\mathbf{y}) \cdot \mathrm{grad}_{\mathbf{y}} \frac{1}{\|\mathbf{x} - \mathbf{y}\|}$$
$$= \boldsymbol{\omega}(\mathbf{y}) \cdot \mathrm{grad}_{\mathbf{y}} \frac{1}{\|\mathbf{x} - \mathbf{y}\|}$$

because $\boldsymbol{\omega}$ is divergence-free according to (1.18)

Therefore, we have

$$\mathrm{div}_{\mathbf{x}} \frac{\boldsymbol{\omega}(\mathbf{y})}{\|\mathbf{x} - \mathbf{y}\|} = -\mathrm{div}_{\mathbf{y}} \frac{\boldsymbol{\omega}(\mathbf{y})}{\|\mathbf{x} - \mathbf{y}\|}$$

This makes it possible to write (A.60) in the form

$$\mathrm{div}\,\mathbf{v} = -\frac{1}{4\pi} \int_{\mathbb{R}^3} \mathrm{div}_{\mathbf{y}} \frac{\boldsymbol{\omega}(\mathbf{y})}{\|\mathbf{x} - \mathbf{y}\|}\, dV_{\mathbf{y}}$$

Applying Gauss' integral theorem, we get

$$\int_{\|\mathbf{x}-\mathbf{y}\|\leq R} \mathrm{div}_{\mathbf{y}} \frac{\boldsymbol{\omega}(\mathbf{y})}{\|\mathbf{x} - \mathbf{y}\|}\, dV_{\mathbf{y}} = \int_{\|\mathbf{x}-\mathbf{y}\|=R} \frac{\boldsymbol{\omega}(\mathbf{y})}{\|\mathbf{x} - \mathbf{y}\|} \cdot \mathbf{n}\, dS_{\mathbf{y}}$$

From

$$\left| \int_{\|\mathbf{x}-\mathbf{y}\|=R} \frac{\omega(\mathbf{y})}{\|\mathbf{x}-\mathbf{y}\|} \cdot \mathbf{n}\, dS_{\mathbf{y}} \right| \le \int_{\|\mathbf{x}-\mathbf{y}\|=R} C R^{1-N}\, dS_{\mathbf{y}} \to 0 \quad \text{for } R \to \infty$$

it follows that $\operatorname{div} \mathbf{v} = 0$ and (A.59) reduces to $\operatorname{rot}(\operatorname{rot} \mathbf{v}) = \omega$. Thus, $\mathbf{u} := \operatorname{rot} \mathbf{v}$ is a velocity field for ω. To prove its uniqueness, we assume another velocity field $\tilde{\mathbf{u}}$ that vanishes at infinity and is divergence-free. Since $\operatorname{rot} \tilde{\mathbf{u}} = \omega$, we get

$$\operatorname{rot}(\mathbf{u} - \tilde{\mathbf{u}}) = 0$$

Lemma 1.1 ensures the existence of a scalar potential ϕ such that

$$\mathbf{u} - \tilde{\mathbf{u}} = \nabla \phi$$

Both fields are divergence-free. Therefore,

$$0 = \operatorname{div}(\mathbf{u} - \tilde{\mathbf{u}}) = \Delta \phi$$

and ϕ is a harmonic function. Because

$$\Delta \frac{\partial \phi}{\partial x_k} = \sum_{i=1}^{3} \frac{\partial^2}{\partial x_i^2} \frac{\partial \phi}{\partial x_k} = \frac{\partial}{\partial x_k} \sum_{i=1}^{3} \frac{\partial^2 \phi}{\partial x_i^2} = 0$$

each partial derivative of ϕ is also harmonic. Furthermore, due to the boundedness of both velocity fields, each $\partial \phi / \partial x_k$ is bounded on $\mathbb{R}^3$. According to Liouville's theorem, the functions $\partial \phi / \partial x_k$ are constant, so that the velocity fields differ only by a constant. Since both fields vanish at infinity, we have shown their equality. $\qquad \square$

References and Recommendations for further reading

1. Vladimirov W.S.: *Equations of Mathematical Physics*. Mir Publishers, Moscow (1984)

A.10 Vorticity in the Frictionless Model

We show an alternative proof for Corollary 5.1. Consider (1.17) for $\mathbf{a} = \mathbf{b} = \mathbf{u}$. Then

$$\frac{1}{2} \operatorname{grad}(\mathbf{u} \cdot \mathbf{u}) = (\mathbf{u} \cdot \nabla)\mathbf{u} + \mathbf{u} \times \operatorname{rot} \mathbf{u}$$

$$(\mathbf{u} \cdot \nabla)\mathbf{u} = \frac{1}{2} \operatorname{grad}(\mathbf{u} \cdot \mathbf{u}) - \mathbf{u} \times \operatorname{rot} \mathbf{u}$$

Substituting this expression into the Euler equation (5.1) yields

$$\frac{\partial \mathbf{u}}{\partial t} + \left[\frac{1}{2}\,\text{grad}\,(\mathbf{u} \cdot \mathbf{u}) - \mathbf{u} \times \text{rot}\,\mathbf{u} \right] + \frac{1}{\varrho}\,\text{grad}\,p = 0$$

$$\frac{\partial}{\partial t}\,\text{rot}\,\mathbf{u} + \text{rot}\left[\frac{1}{2}\,\text{grad}\,(\mathbf{u} \cdot \mathbf{u}) - \mathbf{u} \times \text{rot}\,\mathbf{u} \right] + \frac{1}{\varrho}\,\text{rot}\,\text{grad}\,p = 0$$

$$\frac{\partial}{\partial t}\,\text{rot}\,\mathbf{u} + \text{rot}\left[-\mathbf{u} \times \text{rot}\,\mathbf{u} \right] = 0 \quad \text{by (1.19)}$$

By replacing $\boldsymbol{\omega} = \text{rot}\,\mathbf{u}$, the equation simplifies to

$$\frac{\partial \boldsymbol{\omega}}{\partial t} - \text{rot}\,[\mathbf{u} \times \boldsymbol{\omega}] = 0, \qquad \boldsymbol{\omega}(\mathbf{x}, 0) = 0 \tag{A.61}$$

We want to prove that this flow remains vortex-free. From the linearity of the vector product and the curl operator, it follows that the term $\text{rot}\,[\mathbf{u} \times \boldsymbol{\omega}]$ is linear in $\boldsymbol{\omega}$. Consider $\mathbf{u}(\mathbf{x}, t)$ as given. Then, (A.61) defines for each fixed spatial point $\mathbf{x}$ a linear, non-autonomous system of differential equations

$$\dot{\boldsymbol{\zeta}} = A(t)\,\boldsymbol{\zeta}, \qquad \boldsymbol{\zeta}(0) = 0,$$

whose matrix $A(t)$ depends on $\mathbf{u}(\mathbf{x}, t)$ and its first derivative, i.e. on the velocity and acceleration field. For $z(t) = \|\boldsymbol{\zeta}(t)\| \geq 0$ we obtain the inequality

$$\dot{z} \leq \|A(t)\|\,z, \qquad z(0) = 0,$$

where $\|A(t)\|$ denotes a matrix norm that is compatible with the vector norm. Applying Gronwall's lemma (see [1]) and and considering $z(0) = 0$, it follows that

$$z(t) \leq e^{\int_0^t \|A(\tau)\|\,d\tau}\,z(0) = 0$$

The flow thus remains vortex-free for all times.

References and Further Reading

1. Amann H.: *Ordinary differential equations. An introduction to nonlinear analysis.* De Gruyter, Berlin (1990)

A.11   Numerical Data for the Boundary Layer Equation

For a description of the shooting method for determining the boundary layer thickness, we refer to [1, 2]. A source code for MAPLE is also provided there.

Table A.1 Numerical solution of (5.38) with boundary conditions (5.39), (5.40), (5.41).

η	f''	f'	f
0	0.33206	0	0
0.1	0.33205	0.03321	0.00166
0.2	0.33198	0.06641	0.00664
0.3	0.33181	0.0996	0.01494
0.4	0.33147	0.13276	0.02656
0.5	0.33091	0.16589	0.04149
0.6	0.33008	0.19894	0.05973
0.8	0.32739	0.26471	0.10611
1	0.32301	0.32978	0.16557
1.2	0.31659	0.39378	0.23795
1.4	0.30787	0.45626	0.32298
1.6	0.29666	0.51676	0.42032
1.8	0.28293	0.57476	0.52952
2	0.26675	0.62977	0.65002
2.2	0.24835	0.68131	0.78119
2.4	0.22809	0.72898	0.92229
2.6	0.20645	0.77245	1.0725
2.8	0.18401	0.81151	1.23098
3	0.16136	0.84604	1.39681
3.5	0.10777	0.91304	1.8377
4	0.06423	0.95552	2.30574
4.5	0.03398	0.97951	2.79013
5	0.01591	0.99154	3.28327
5.5	0.00658	0.99688	3.78057
6	0.0024	0.99897	4.27962
6.5	0.00077	0.9997	4.77932
7	0.00022	0.99992	5.27923
8	0.00001	1	6.27921
9	0	1	7.27921
10	0	1	8.27921

References and Further Reading

1. Meade D.B., Haran B.S., White R.E.: *The Shooting Technique for the Solution of Two-Point Boundary Value Problems.* MapleTech 3 (1996), S.85–93
2. Sun B.: *Solving Prandtl-Blasius Boundary Layer Equation Using Maple.* Preprints (2020) 2020080296. https://doi.org/10.20944/preprints202008.0296.v2   Accessed 22. Sept 2024

A.12   Measurement Data for the NACA 64-210 Profile

See Fig. A.6.

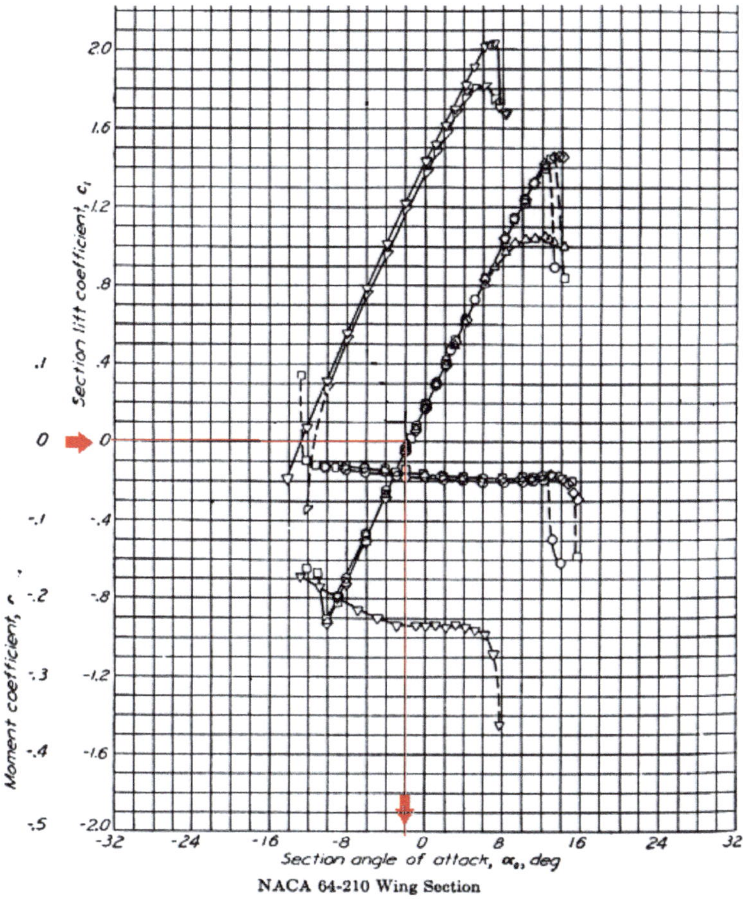

Fig. A.6 Measurement data for $C_l = C_l(\alpha)$ for the NACA 64-210 profile according to Abbott [1], p. 564. If you follow the arrow on the y-axis for $C_l = 0$, you read the value $\alpha \approx -1.8°$ on the x-axis. Image rights: McGraw Hill.

References and Further Reading

1. Abbott I.H., von Doenhoff A.E.: *Theory of Wing Sections. Including a Summary of Airfoil Data.* Dover, New York (1959)

Solutions

To Chap. 1

1.1 We want to prove that $\epsilon_{ij1}\,\epsilon_{1lm} + \epsilon_{ij2}\,\epsilon_{2lm} + \epsilon_{ij3}\,\epsilon_{3lm} = \delta_{il}\,\delta_{jm} - \delta_{im}\,\delta_{jl}$.

Case 1: Let $i = j$ or $l = m$. Then $\epsilon_{ijk}\,\epsilon_{klm} = 0$ and

$$\delta_{il}\,\delta_{jm} - \delta_{im}\,\delta_{jl} = \delta_{il}\,\delta_{im} - \delta_{im}\,\delta_{il} = 0$$

Case 2: Let $i \neq j$ and $l \neq m$. There remains exactly one value $k \notin \{i, j\}$ and $k \notin \{l, m\}$.

Case 2a: For $i = l$ and $j = m$, it follows that $\delta_{il} = \delta_{jm} = 1$ and $\delta_{im} = \delta_{jl} = 0$. Thus,

$$\delta_{il}\,\delta_{jm} - \delta_{im}\,\delta_{jl} = 1$$

The permutations (ijk) and (kij) have the same order, i.e. both are either even or odd. Thus, $\epsilon_{ijk}\,\epsilon_{klm} = \epsilon_{ijk}\,\epsilon_{kij} = 1$.

Case 2b: For $i = m$ and $j = l$, it follows that $\delta_{il}\,\delta_{jm} = 0$ and $\delta_{im}\,\delta_{jl} = 1$. Thus,

$$\delta_{il}\,\delta_{jm} - \delta_{im}\,\delta_{jl} = -1$$

(ijk) and (kji) have different orders. Therefore, $\epsilon_{ijk}\,\epsilon_{klm} = \epsilon_{ijk}\,\epsilon_{kji} = -1$.

1.2 The k-fold of the second line should be added to the first line. Let

$$D = \begin{bmatrix} a_x & a_y & a_z \\ b_x & b_y & b_z \\ c_x & c_y & c_z \end{bmatrix}, \qquad D_k = \begin{bmatrix} a_x + kb_x & a_y + kb_y & a_z + kb_z \\ b_x & b_y & b_z \\ c_x & c_y & c_z \end{bmatrix}$$

Then

$$\det D_k = (\mathbf{a} + k\mathbf{b}) \cdot (\mathbf{b} \times \mathbf{c}) = \mathbf{a} \cdot (\mathbf{b} \times \mathbf{c}) + k\mathbf{b} \cdot (\mathbf{b} \times \mathbf{c})$$

$$= \mathbf{a} \cdot (\mathbf{b} \times \mathbf{c}) + k\mathbf{c} \cdot \underbrace{(\mathbf{b} \times \mathbf{b})}_{=0} = \det D$$

For all other lines, the statement follows after the corresponding permutation of the lines.

1.3

$$\det \begin{bmatrix} ka_x & ka_y & ka_z \\ b_x & b_y & b_z \\ c_x & c_y & c_z \end{bmatrix} = (k\mathbf{a}) \cdot (\mathbf{b} \times \mathbf{c}) = k[\mathbf{a} \cdot (\mathbf{b} \times \mathbf{c})] = k \det \begin{bmatrix} a_x & a_y & a_z \\ b_x & b_y & b_z \\ c_x & c_y & c_z \end{bmatrix}$$

1.4 Proof of (1.13):

$$\mathrm{rot}\,(b\,\mathbf{a}) = \epsilon_{ijk}\,\partial_i(ba_j) = \epsilon_{ijk}\,(\partial_i b)a_j + b\epsilon_{ijk}\,\partial_i a_j = (\mathrm{grad}\,b) \times \mathbf{a} + b\,\mathrm{rot}\,\mathbf{a}$$

Proof of (1.14):

$$\mathrm{rot}\,(\mathrm{rot}\,\mathbf{a}) = \epsilon_{lkn}\,\partial_l(\epsilon_{ijk}\,\partial_i a_j) = \epsilon_{lkn}\epsilon_{ijk}\,\partial_l\partial_i a_j = -\epsilon_{kln}\epsilon_{kij}\,\partial_l\partial_i a_j$$

$$= (\delta_{lj}\,\delta_{ni} - \delta_{li}\,\delta_{nj})\,\partial_l\partial_i a_j = \delta_{lj}\,\delta_{ni}\,\partial_l\partial_i a_j - \delta_{li}\,\delta_{nj}\,\partial_l\partial_i a_j$$

$$= (\partial_j\partial_n a_j) - (\partial_i\partial_i a_j) = \mathrm{grad}\,\mathrm{div}\,\mathbf{a} - \Delta\mathbf{a}$$

Proof of (1.15):

$$\mathrm{div}\,(b\,\mathbf{a}) = \partial_i(ba_i) = b\,\partial_i a_i + a_i\,(\partial_i b) = b\,\mathrm{div}\,\mathbf{a} + \mathbf{a} \cdot \mathrm{grad}\,b$$

Proof of (1.16):

$$\mathrm{rot}\,(\mathbf{a} \times \mathbf{b}) = \epsilon_{nkm}\,\partial_n(\epsilon_{ijk}a_i b_j) = \epsilon_{nkm}\epsilon_{ijk}\,\partial_n(a_i b_j) = -\epsilon_{knm}\epsilon_{kij}\,\partial_n(a_i b_j)$$

$$= (\delta_{nj}\,\delta_{mi} - \delta_{ni}\,\delta_{mj})\,\partial_n(a_i b_j) = \delta_{nj}\,\delta_{mi}\,\partial_n(a_i b_j) - \delta_{ni}\,\delta_{mj}\,\partial_n(a_i b_j)$$

$$= \partial_n(a_m b_n) - \partial_n(a_n b_m) = b_n\,\partial_n a_m + a_m\,(\partial_n b_n) - (a_n\,\partial_n b_m + b_m\,\partial_n a_n)$$

$$= (\mathbf{b} \cdot \nabla)\mathbf{a} - (\mathbf{a} \cdot \nabla)\mathbf{b} + \mathbf{a}\,\mathrm{div}\,\mathbf{b} - \mathbf{b}\,\mathrm{div}\,\mathbf{a}$$

Proof of (1.17):

$$\mathbf{a} \times (\mathrm{rot}\,\mathbf{b}) = \epsilon_{lkm}a_l(\epsilon_{ijk}\,\partial_i b_j) = -\epsilon_{klm}\,\epsilon_{kij}a_l\,\partial_i b_j$$

$$= (\delta_{lj}\,\delta_{mi} - \delta_{li}\,\delta_{mj})\,a_l\,\partial_i b_j = \delta_{lj}\delta_{mi}a_l\,\partial_i b_j - \delta_{li}\delta_{mj}a_l\,\partial_i b_j$$

$$= (a_j\,\partial_i b_j) - (a_i\,\partial_i b_j)$$

and similarly

$$\mathbf{b} \times (\operatorname{rot} \mathbf{a}) = (b_j\, \partial_i a_j) - (b_i\, \partial_i a_j)$$

Therefore,

$$(\mathbf{a} \cdot \nabla)\,\mathbf{b} + \mathbf{a} \times (\operatorname{rot} \mathbf{b}) = a_j\, \partial_j b_i + a_j\, \partial_i b_j - a_i\, \partial_i b_j = (a_j\, \partial_i b_j)$$

$$(\mathbf{b} \cdot \nabla)\,\mathbf{a} + \mathbf{b} \times (\operatorname{rot} \mathbf{a}) = b_j\, \partial_j a_i + b_j\, \partial_i a_j - b_i\, \partial_i a_j = (b_j\, \partial_i a_j)$$

and we obtain

$$(\mathbf{a} \cdot \nabla)\,\mathbf{b} + \mathbf{a} \times (\operatorname{rot} \mathbf{b}) + (\mathbf{b} \cdot \nabla)\,\mathbf{a} + \mathbf{b} \times (\operatorname{rot} \mathbf{a}) = a_j\, \partial_i b_j + b_j\, \partial_i a_j$$
$$= (\partial_i (a_j b_j))$$
$$= \operatorname{grad} (\mathbf{a} \cdot \mathbf{b})$$

Proof of (1.18): We have $\operatorname{div} \operatorname{rot} \mathbf{a} = \partial_k(\epsilon_{ijk}\, \partial_i a_j) = \epsilon_{ijk}\, \partial_k \partial_i a_j$. In the sum, all terms with repeating indices disappear. Thus, (ijk) must be a permutation of (123). By swapping k and i, each such index triplet appears twice. From $\epsilon_{ijk} = -\epsilon_{kji}$, it follows that $\epsilon_{ijk}\, \partial_k \partial_i a_j = 0$.

Proof of (1.19): We have $\operatorname{rot} \operatorname{grad} b = (\epsilon_{ijk}\, \partial_i \partial_j b)$. Again, all triplets with repeating indices disappear and all index pairs $i \neq j$ appear twice. From $\epsilon_{jik} = -\epsilon_{ijk}$ we obtain the assertion.

1.5 Using (1.8), we find that

$$\operatorname{rot} \mathbf{a} = \det \begin{bmatrix} \mathbf{e}_x & \mathbf{e}_y & \mathbf{e}_z \\ \partial_x & \partial_y & \partial_z \\ u & v & 0 \end{bmatrix} = -\mathbf{e}_x \partial_z v - \mathbf{e}_y \partial_z u + \mathbf{e}_z(\partial_x v - \partial_y u) = \mathbf{e}_z(\partial_x v - \partial_y u),$$

(In the last step, we made use of the fact that u, v are independent of z.)

1.6 Consider the tangentially oriented differential $d\mathbf{x} = \begin{pmatrix} dx \\ dy \end{pmatrix}$.

Since $\begin{pmatrix} dx \\ dy \end{pmatrix} \cdot \begin{pmatrix} dy \\ -dx \end{pmatrix} = 0$, it is perpendicular to $\mathbf{n}\, ds = \begin{pmatrix} dy \\ -dx \end{pmatrix}$. Thus, the latter is a normally directed differential. The same applies to $-\mathbf{n}\, ds = \begin{pmatrix} -dy \\ dx \end{pmatrix}$.

Therefore, we obtain

$$\mathbf{u} \cdot \mathbf{n}\, ds = \begin{pmatrix} u \\ v \end{pmatrix} \cdot \begin{pmatrix} dy \\ -dx \end{pmatrix} = u\, dy - v\, dx$$

1.7 Applying Gauss's integral theorem, it follows that

$$\int_\Sigma \mathbf{n}\,dS = \int_{\text{int}\Sigma} (\nabla 1)\,dV = 0$$

Furthermore, the result already follows from symmetry considerations.

1.8

(a) Using (1.21), it follows that

$$\text{rot}\,\mathbf{u} = \frac{\partial}{\partial x}\left(\frac{x}{x^2 + y^2}\right) + \frac{\partial}{\partial y}\left(\frac{y}{x^2 + y^2}\right) = 0$$

(b) When integrating along the unit circle, we obtain

$$\oint \mathbf{u}\,d\mathbf{x} = \int (-y\,dx + x\,dy) = \int_{\phi=0}^{2\pi} d\phi = 2\pi \neq 0$$

(The second integral is evaluated in polar coordinates $x = r\cos\phi$, $y = r\sin\phi$ for $r = 1$, where we have $dx = -r\sin\phi\,d\phi$, $dy = r\cos\phi\,d\phi$.)

1.9 Proof of (1.28): For $r = \left(\sum(y_k - x_k)^2\right)^{\frac{1}{2}}$ it follows that

$$\frac{\partial r}{\partial y_i} = \frac{1}{2}\left(\sum[y_k - x_k]^2\right)^{-\frac{1}{2}} 2(y_i - x_i) = \frac{y_i - x_i}{r}$$

and therefore (1.28).
Proof of (1.29): Using (1.28) twice, it follows that

$$\mathbf{e}_r = \nabla_y r = \frac{\mathbf{y} - \mathbf{x}}{r} = -\frac{\mathbf{x} - \mathbf{y}}{r} = -\nabla_x r$$

Proof of (1.30): The statement follows from the chain rule and (1.29).

1.10 Using (1.29) and (1.30) for $f(r) = 1/r$, we get

$$\frac{dG(\mathbf{x}, \mathbf{y})}{d\mathbf{n}} = \mathbf{n} \cdot \nabla_y\left(-\frac{1}{4\pi r}\right) = \frac{1}{4\pi}\mathbf{n} \cdot \nabla_x\left(\frac{1}{r}\right) = \frac{1}{4\pi}r^{-2}\mathbf{n} \cdot \mathbf{e}_r$$

To Chap. 2

2.1

$$\frac{D(f\,g)}{Dt} = \frac{\partial(f\,g)}{\partial t} + \mathbf{u} \cdot \nabla(f\,g) = f\frac{\partial g}{\partial t} + g\frac{\partial f}{\partial t} + \mathbf{u} \cdot [f\,\nabla g + g\,\nabla f] = f\frac{Dg}{Dt} + g\frac{Df}{Dt}$$

2.2 Let $C_1(A, P)$ and $C_2(P, A)$ be two different paths which together form a closed curve C and enclose the set $\Sigma \subset \mathbb{R}^2$. Using Theorem 1.1 and $\operatorname{div} \mathbf{u} = 0$, we obtain

$$\int_{C_1(A,P)} \mathbf{u} \cdot \mathbf{n}\, ds + \int_{C_2(P,A)} \mathbf{u} \cdot \mathbf{n}\, ds = \int_C \mathbf{u} \cdot \mathbf{n}\, ds = \int_\Sigma \operatorname{div} \mathbf{u}\, dA = 0$$

and therefore $\int_{C_1(A,P)} \mathbf{u} \cdot \mathbf{n}\, ds = \int_{C_2(A,P)} \mathbf{u} \cdot \mathbf{n}\, ds$.

2.3 For two arbitrary points $P_1 = \begin{pmatrix} x_1 \\ y_1 \end{pmatrix}$ and $P_2 = \begin{pmatrix} x_2 \\ y_2 \end{pmatrix}$ on the same streamline, we want to show $\psi(x_1, y_1) = \psi(x_2, y_2)$. First, we choose a path $C_1(A, P_1)$. Then, we construct $C_2(A, P_2) = C_1(A, P_1) \cup C(P_1, P_2)$, where the curve $C(P_1, P_2)$ lies on the common streamline of P_1 and P_2. Since the velocity field $\mathbf{u}$ is tangential to the streamlines, it is

$$\mathbf{u} \cdot \mathbf{n} = 0 \qquad \text{on } C(P_1, P_2)$$

and consequently

$$\int_{C_2(A,P_2)} \mathbf{u} \cdot \mathbf{n}\, ds = \int_{C_1(A,P_1)} \mathbf{u} \cdot \mathbf{n}\, ds + \int_{C(P_1,P_2)} \mathbf{u} \cdot \mathbf{n}\, ds = \int_{C_1(A,P_1)} \mathbf{u} \cdot \mathbf{n}\, ds$$

To Chap. 3

3.1 The shear force $\mathbf{S}_i$ is given by

$$\mathbf{S}_i = \mathbf{S}\mathbf{e}_i = \begin{bmatrix} 0 & \sigma_{12} & \sigma_{13} \\ \sigma_{21} & 0 & \sigma_{23} \\ \sigma_{31} & \sigma_{32} & 0 \end{bmatrix} \mathbf{e}_i = \begin{pmatrix} \sigma_{1i} \\ \sigma_{2i} \\ \sigma_{3i} \end{pmatrix}$$

Because of (3.8), it holds $\sigma_{ii} = 0$, i.e. the shear force contains no component in the direction $\mathbf{e}_i$.

3.2 We calculate the derivatives using the chain rule, i.e.

$$\frac{\partial u}{\partial t} = \frac{\partial (U u')}{\partial t} = U \frac{\partial u'}{\partial t'} \frac{\partial t'}{\partial t} = U \frac{\partial u'}{\partial t'} \frac{U}{L} = \frac{U^2}{L} \frac{\partial u'}{\partial t'}$$

$$\frac{\partial p}{\partial x} = \frac{\partial (\varrho U^2 p')}{\partial x} = \varrho U^2 \frac{\partial p'}{\partial x'} \frac{\partial x'}{\partial x} = \varrho U^2 \frac{\partial p'}{\partial x'} \frac{1}{L} = \frac{\varrho U^2}{L} \frac{\partial p'}{\partial x'}$$

By substitution we obtain (3.24), (3.25).

3.3 Using (1.8) and (1.10), we get

$$\text{rot } \Delta\mathbf{u} = \epsilon_{ijk}\, \partial_i(\partial_l\partial_l a_j) = \partial_l\partial_l(\epsilon_{ijk}\, \partial_i a_j) = \Delta\boldsymbol{\omega}$$

To Chap. 4

4.1 The streamlines create a suction that causes the umbrella to fold over. In the new shape, the streamlines change so that the direction of suction is reversed. The umbrella folds back again.

4.2 According to Bernoulli's theorem (4.6), it holds that

$$\frac{\partial p}{\partial s} = -2\varrho\,\frac{\partial \|\mathbf{u}\|^2}{\partial s} \qquad \text{along a streamline}$$

4.3 A vortex-free flow with constant density is divergence-free by Corollary 2.1 and incompressible by Corollary 2.3. Lemma 3.4 implies that it is inviscid. This applies in both two- and three-dimensional domains.

4.4 The assertion follows from Lemma 1.1, Statement (2).

4.5 For $\mathbf{u} = \text{grad }\phi = (\partial_j\phi)$, we have

$$\Delta\mathbf{u} = \partial_i\partial_i u_j = \partial_i\partial_i(\partial_j\phi) = \partial_j\partial_i(\partial_i\phi) = \partial_j\underbrace{(\partial_i u_i)}_{=0} = 0 \quad \text{due to incompressibility}$$

Thus, the term $\nu\Delta\mathbf{u}$ does not contribute in the Navier-Stokes equation (3.20) and we get its inviscid-flow approximation (3.30).

4.6 According to Problem 4.5, the flow is inviscid. Thus, the Navier-Stokes equation (3.20) simplifies to (3.30). Therefore,

$$\frac{\partial\mathbf{u}}{\partial t} + (\mathbf{u}\cdot\nabla)\mathbf{u} + \nabla\!\left(\frac{p}{\varrho}\right) + \nabla\Phi = 0$$

Because of (1.19), it holds $\text{rot }\mathbf{u} = \text{rot grad }\phi = 0$. As in the proof of Theorem 3.2, we use the identity (3.36), which simplifies in the vortex-free case to

$$(\mathbf{u}\cdot\nabla)\,\mathbf{u} = \frac{1}{2}\nabla\,(\|\mathbf{u}\|^2)$$

From $\mathbf{u} = \text{grad }\phi$, it follows that

$$\nabla\left(\frac{\partial\phi}{\partial t} + \frac{1}{2}(\|\text{grad }\phi\|^2) + \frac{p}{\varrho} + \Phi\right) = 0 \qquad \text{in the entire domain,}$$

which proves the assertion.

4.7 We decompose the boundary $C = C_1 \cup C_2$ into the upper path C_1 and the lower path C_2, whose lengths are approximately equal. Then we have

$$\Gamma = \int_{C_1} \mathbf{u}\,d\mathbf{x} + \int_{C_2} \mathbf{u}\,d\mathbf{x}$$

Since C is traversed in the positive direction, we see that

$$\mathbf{u}\,d\mathbf{x} < 0 \text{ on } C_1 \quad \text{and} \quad \mathbf{u}\,d\mathbf{x} > 0 \text{ on } C_2$$

We denote the magnitudes of the average flow velocities on C_1 and C_2 by u_1 and u_2, respectively. From Fig. 4.12, we can see that $u_1 > u_2$. Thus,

$$\Gamma \approx -u_1 \int_{C_1} ds + u_2 \int_{C_2} ds \approx (u_2 - u_1) \int_{C_1} ds < 0$$

4.8 The fluid cannot penetrate the boundary. Therefore, the stream function ψ remains constant there. See (2.10).

4.9 Using $z - z_0 = r e^{i\theta}$, we get from (4.16)

$$w(z) = -\frac{\Gamma\theta}{2\pi} + i\frac{\Gamma \ln r}{2\pi}$$

The streamlines

$$\psi = \Gamma \ln r / 2\pi = \text{const}$$

form circles $r = \text{const}$ around z_0, while the level sets of the velocity potential

$$\phi = -\Gamma\theta / 2\pi = \text{const}$$

correspond to the rays $\theta = \text{const}$.

4.10 For $w_1 = \phi_1 + i\psi_1$, $w_2 = \phi_2 + i\psi_2$, the stream functions are constant on the boundary, i.e. we have $\psi_1 = c_1$, $\psi_2 = c_2$. For

$$a_1 w_1 + a_2 w_2 = a_1 \phi_1 + a_2 \phi_2 + i(a_1 \psi_1 + a_2 \psi_2)$$

we get $a_1 \psi_1 + a_2 \psi_2 = a_1 c_1 + a_2 c_2$ on the boundary.

4.11 We calculate the complex velocity

$$\mathcal{U}(\zeta) = \frac{dw}{d\zeta} = u_\infty \left[e^{i\alpha} - \frac{a^2 e^{-i\alpha}}{(\zeta - \zeta_c)^2} \right] - \frac{i\Gamma}{2\pi(\zeta - \zeta_c)}$$

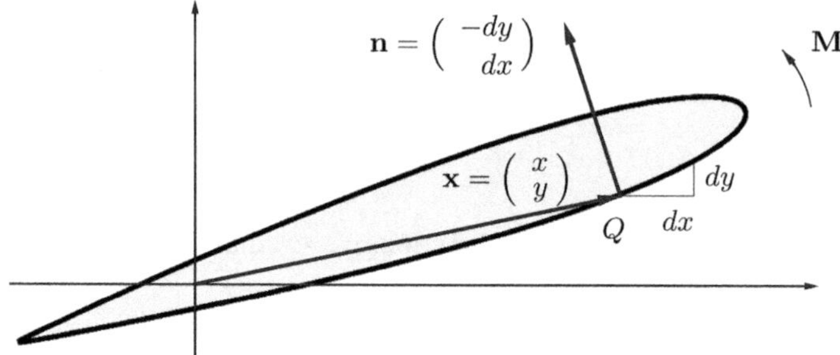

Fig. A.7 The pressure force $p\mathbf{n}$ at $Q = (x, y)$ generates a torque $\mathrm{d}M$ around the origin.

Then, $\lim_{\zeta \to \infty} \mathcal{U}(\zeta) = u_\infty e^{i\alpha}$.

4.12 From (4.12) and (4.20) it follows that

$$
\oint_{K(C,a)} \mathcal{U} \,\mathrm{d}\zeta = u_\infty \left[e^{i\alpha} \oint_{K(C,a)} \mathrm{d}\zeta - a^2 e^{-i\alpha} \oint_{K(C,a)} \frac{\mathrm{d}\zeta}{(\zeta - \zeta_c)^2} \right]
$$

$$
- \frac{i\Gamma}{2\pi} \oint_{K(C,a)} \frac{\mathrm{d}\zeta}{\zeta - \zeta_c}
$$

$$
= \Gamma \qquad \text{by Theorem A.1 and Problem A.8}
$$

4.13 See Fig. A.7. If we supplement all vectors with a z-component, then

$$
\mathrm{d}\mathbf{M} = \begin{pmatrix} x \\ y \\ 0 \end{pmatrix} \times p \begin{pmatrix} -\mathrm{d}y \\ \mathrm{d}x \\ 0 \end{pmatrix} = \begin{pmatrix} 0 \\ 0 \\ p(x\,\mathrm{d}x + y\,\mathrm{d}y) \end{pmatrix}
$$

4.14

$$
\frac{\mathrm{d}\bar{w}}{\mathrm{d}\bar{z}} = \lim_{\Delta\bar{z}\to 0} \frac{\Delta\bar{w}}{\Delta\bar{z}} = \lim_{\Delta z\to 0} \overline{\left(\frac{\Delta w}{\Delta z} \right)} = \overline{\left(\frac{\mathrm{d}w}{\mathrm{d}z} \right)}
$$

4.15 Since C forms a streamline, it follows from (2.10) that

$$
\psi = \text{const} \qquad \text{on } C
$$

From $\mathrm{d}w = \mathrm{d}\phi + i\mathrm{d}\psi$ and $\mathrm{d}\bar{w} = \mathrm{d}\phi - i\mathrm{d}\psi$, we get the claim.

4.16 Using the addition theorem, we get from (4.35) that

$$
x + iy = c \cosh(\xi + i\eta) = c \cosh\xi \cos\eta + ic \sinh\xi \sin\eta
$$

Therefore, $x = c \cosh\xi \cos\eta$, $y = c \sinh\xi \sin\eta$, which implies (4.36), (4.37).

4.17 For the boundary points $\zeta = \xi_0 + i\eta$, it holds that

$$w(\xi_0 + i\eta) = cA \cosh(i(\eta - \alpha)) = \frac{cA}{2}(e^{i(\eta-\alpha)} + e^{-i(\eta-\alpha)}) = cA \cos(\eta - \alpha)$$

Thus, $\operatorname{Im} w = 0$ on the boundary.

4.18 From (4.42) we get

$$\begin{aligned}
\mathcal{U}_\infty = \lim_{z \to \infty} \frac{d\hat{w}}{dz} &= A \lim_{\zeta \to \infty} \left[\cosh \zeta_0 - \frac{\sinh \zeta_0 \cosh \zeta}{\sinh \zeta} \right] \\
&= A[\cosh \zeta_0 - \sinh \zeta_0] = Ae^{-\zeta_0} \\
&= Ae^{-\xi_0}(\cos \alpha - i \sin \alpha)
\end{aligned}$$

Using the transformation (4.11), we obtain (4.45).

4.19 From the condition $\mathcal{U} = 0$ and (4.41), we get $dw/d\zeta = 0$. Using (4.40), it follows that $\sinh(\zeta - \xi_0 - i\alpha) = 0$. For $\zeta = \xi_0 + i\eta$, we get

$$i(\eta - \alpha) = 0 \qquad \text{or} \qquad i(\eta - \alpha) = i\pi,$$

from which the sought angles η result.

4.20 Taking into account $\sin(2\alpha) = 2 \sin \alpha \cos \alpha$, it follows from (4.47) that

$$M = -k \sin(2\alpha) \qquad \text{with } k > 0$$

Then, $M = 0$ for $\alpha = 0$ and $\alpha = \pi/2$. Using the approach $\alpha = \pi/2 + \epsilon$, we examine a small deviation from the transverse position. Since

$$\sin(2\alpha) = \sin(\pi + 2\epsilon) = -\sin(2\epsilon),$$

we find that $M = k \sin(2\epsilon)$. Thus, ϵ and M have the same sign. For $\epsilon > 0$ (deflection of the ship's hull counterclockwise), we get $M > 0$ (clockwise) and the torque causes a return to the initial position. The same counteracting effect occurs for $\epsilon < 0$ and $M < 0$. In the longitudinal position, we set $\alpha = 0 + \epsilon$ and obtain $M = -k \sin(2\epsilon)$. A small deflection of the ship's hull causes an amplifying torque.

4.21

(a) For points on the unit circle $\zeta = e^{i\phi}$, it holds $z = e^{i\phi} + e^{-i\phi} = 2 \cos \phi$. Thus, the unit circle $\phi \in [0, 2\pi]$ is mapped to the interval $[-2, 2]$.

(b) From (4.48) we get $0 = \zeta^2 - \zeta z + 1$ and $\zeta = \frac{z}{2} \pm \sqrt{\frac{z^2}{4} - 1}$. Therefore, each $z \neq \pm 2$ is assigned to two values of ζ. Let us consider (4.48) for a given $z \neq \pm 2$. If ζ is a solution, then $\frac{1}{\zeta}$ is the second solution. $|\zeta| > 1$ holds if and only if

$|\frac{1}{\zeta}| < 1$. Thus, the Joukowski transformation consists of two bijections, the first of which maps the interior of the circle and the second of which maps the exterior to $\mathbb{C} \setminus [-2; 2]$.

4.22 By ζ_0 we denote the centre coordinates of $K(C, a)$. In the complex plane, the circle is described by

$$|\zeta - \zeta_0|^2 = a^2$$

and we have

$$(\zeta - \zeta_0)(\bar{\zeta} - \bar{\zeta}_0) = a^2 \tag{S.1}$$

$$\zeta\bar{\zeta} - \bar{\zeta}_0\zeta - \zeta_0\bar{\zeta} + \zeta_0\bar{\zeta}_0 - a^2 = 0$$

Taking into account (4.49), we substitute $\zeta = 1/\zeta_1$, $\bar{\zeta} = 1/\bar{\zeta}_1$ and get

$$\frac{1}{\zeta_1}\frac{1}{\bar{\zeta}_1} - \bar{\zeta}_0\frac{1}{\zeta_1} - \zeta_0\frac{1}{\bar{\zeta}_1} + \zeta_0\bar{\zeta}_0 - a^2 = 0$$

$$1 - \bar{\zeta}_0\bar{\zeta}_1 - \zeta_0\zeta_1 - (a^2 - \zeta_0\bar{\zeta}_0)\zeta_1\bar{\zeta}_1 = 0$$

Division by $\eta = a^2 - \zeta_0\bar{\zeta}_0 > 0$ gives

$$\zeta_1\bar{\zeta}_1 + \frac{\bar{\zeta}_0}{\eta}\bar{\zeta}_1 + \frac{\zeta_0}{\eta}\zeta_1 - \frac{1}{\eta} = 0$$

$$\zeta_1\bar{\zeta}_1 + \frac{\bar{\zeta}_0}{\eta}\bar{\zeta}_1 + \frac{\zeta_0}{\eta}\zeta_1 + \frac{\zeta_0\bar{\zeta}_0}{\eta^2} = \frac{\zeta_0\bar{\zeta}_0}{\eta^2} + \frac{1}{\eta}$$

$$(\zeta_1 + \frac{\bar{\zeta}_0}{\eta})(\bar{\zeta}_1 + \frac{\zeta_0}{\eta}) = \frac{\zeta_0\bar{\zeta}_0}{\eta^2} + \frac{1}{\eta}$$

By comparison with (S.1), we see that ζ_1 describes a circle with radius $R = \sqrt{\frac{\zeta_0\bar{\zeta}_0}{\eta^2} + \frac{1}{\eta}} = \frac{a}{\eta}$ and the centre coordinates $\tilde{\zeta} = -\frac{\bar{\zeta}_0}{\eta}$.

4.23 For $\zeta \neq 0$, the Joukowski transformation is analytic. Moreover,

$$\frac{dz}{d\zeta} = 1 - \frac{1}{\zeta^2} \neq 0 \qquad \text{for } \zeta \notin \{0, \pm 1\}$$

The claim follows from Problem A.9.

4.24 From (4.48) we get

$$z + 2 = \zeta + 2 + \frac{1}{\zeta} = (\zeta + 1)^2, \quad z - 2 = \zeta - 2 + \frac{1}{\zeta} = (\zeta - 1)^2$$

By dividing both equations, we obtain the statement.

4.25 By Problem 4.21b, $\zeta(z)$ is not uniquely determined. However, according to Fig. 4.20, the interior of the unit circle does not belong to the domain of w. Thus, $\zeta(z)$ becomes a uniquely defined bijection.

4.26 From (4.59) it follows that

$$\lim_{z\to\infty}\hat{\mathcal{U}}(z) = \lim_{\zeta\to\infty}\mathcal{U}(\zeta)\,\frac{1}{1-\frac{1}{\zeta^2}} = u_\infty e^{i\alpha}$$

Thus, α can be interpreted as the angle of attack.

4.27 By $J(K)$ we denote the image of the circle $K(C, a)$ under the Joukowski transformation (see Fig. 4.20). From (4.12) and (4.58), it follows that

$$\Gamma = \oint_{J(K)}\hat{\mathcal{U}}\,dz = \oint_{J(K)}\mathcal{U}\frac{d\zeta}{dz}\,dz = \oint_{K(C,a)}\mathcal{U}\,d\zeta$$

Using (4.57), we obtain

$$\Gamma = u_\infty\left[e^{i\alpha}\oint_{K(C,a)}d\zeta - a^2 e^{-i\alpha}\oint_{K(C,a)}\frac{d\zeta}{(\zeta-\zeta_c)^2} + i\,2a\sin\beta\oint_{K(C,a)}\frac{d\zeta}{\zeta-\zeta_c}\right]$$
$$= -4a\pi u_\infty\sin\beta \qquad \text{(see Problem 4.12)}$$

To Chap. 5

5.1 It holds $\mathbf{u} = (u(y), 0)$, where $u(y)$ is monotonically increasing. Thus, we get

$$\mathrm{rot}\,\mathbf{u} = \frac{\partial v}{\partial x} - \frac{\partial u}{\partial y} = -\frac{\partial u}{\partial y} \neq 0$$

5.2 See Problem 3.2.

5.3 No. Due to the non-vanishing circulation, it contains a bound vortex.

5.4 From (5.46) we get

$$0 = 2\zeta^2 - 4z\zeta + 2$$
$$(z-1)(\zeta+1)^2 = (z+1)(\zeta-1)^2$$
$$\frac{\zeta+1}{\zeta-1} = \sqrt{\frac{z+1}{z-1}}$$

5.5 From $e^{i\theta} = \cos\theta + i\sin\theta$, it can be seen that at $0 \leq \theta \leq \pi$ the upper semicircle is traversed and at $\pi \leq \theta \leq 2\pi$ the lower semicircle is traversed. For $z = \cos\theta$ we get the top and bottom of the plate for the same intervals of angles. It holds

$$\zeta^{-1} - \zeta = e^{-i\theta} - e^{i\theta} = -2i\sin\theta$$

On the plate, i.e. for $z = \cos\theta$ and $-1 \le z \le 1$, we get from $1 = \sin^2\theta + z^2$

$$\sin\theta = \sqrt{1-z^2} = i\sqrt{z^2-1} \qquad \text{for } \sin\theta \ge 0 \quad \text{(top side)}$$
$$\sin\theta = -\sqrt{1-z^2} = -i\sqrt{z^2-1} \quad \text{for } \sin\theta \le 0 \quad \text{(bottom side)}$$

and thus

$$\zeta^{-1} - \zeta = \begin{cases} 2\sqrt{z^2-1} & \text{if } 0 \le \theta \le \pi \\ -2\sqrt{z^2-1} & \text{if } \pi \le \theta \le 2\pi \end{cases}$$

5.6 We surround the infinitesimal segment $d\zeta$ of a vortex sheet by a closed curve C and use Theorem 5.3. Then we convert the integrals into real integrals and apply Stokes' theorem. When the diameter of C tends to zero, we obtain the assertion.

5.7 z_0 describes the position of a free vortex and is real due to W2. Since free vortices drift from the trailing edge $z = 1$ to the right, we have $z_0 > 1$. The inversion of (5.46) yields

$$\zeta_0 = z_0 \pm \sqrt{z_0^2 - 1},$$

i.e. ζ_0 is real. Therefore, ζ_0^{-1} is also real.

5.8 It holds

$$\ln(\zeta - \zeta_0) - \ln(\zeta - \zeta_0^{-1}) = \ln r_1 e^{i\theta_1} - \ln r_2 e^{i\theta_2} = \ln\frac{r_1}{r_2} + i(\theta_1 - \theta_2)$$

By substitution into (5.60) we obtain the statement.

5.9 Using the law of cosines, we calculate r_1 and r_2

$$r_1^2 = 1 + x^2 - 2x\cos\theta \qquad \text{in triangle } OX'Z$$
$$r_2^2 = 1 + \frac{1}{x^2} - \frac{2}{x}\cos\theta \qquad \text{in triangle } OXZ$$

Therefore,

$$\left(\frac{r_1}{r_2}\right)^2 = \frac{1 + x^2 - 2x\cos\theta}{1 + \frac{1}{x^2} - \frac{2}{x}\cos\theta} = x^2$$

5.10 In triangle $X'XZ$ it holds that

$$\sin(\theta_1 - \theta_2) = \frac{x - x^{-1}}{r_2} \sin\theta_1 \qquad \text{(law of sines)} \tag{S.2}$$

$$\cos(\theta_1 - \theta_2) = \frac{r_1^2 + r_2^2 - (x - x^{-1})^2}{2r_1 r_2} \qquad \text{(law of cosines)}$$

$$= \frac{r_1^2 + r_2^2 - x^2 + 2 - (x^{-1})^2}{2r_1 r_2} \tag{S.3}$$

Applying the law of cosines in the triangles $OX'Z$ and OXZ, we get

$$r_2^2 = r^2 + (x^{-1})^2 - 2rx^{-1}\cos\theta$$
$$r_1^2 = r^2 + x^2 - 2rx\cos\theta$$

Adding both equations gives

$$r_1^2 + r_2^2 = 2r^2 + x^2 + (x^{-1})^2 - 2r(x + x^{-1})\cos\theta \tag{S.4}$$

From (S.3) and (S.4), it follows that

$$\cos(\theta_1 - \theta_2) = \frac{2r^2 + 2 - 2r(x + x^{-1})\cos\theta}{2r_1 r_2} \tag{S.5}$$

Applying the law of sines in triangle OXZ, we get

$$r_1 \sin\theta_1 = r \sin\theta \tag{S.6}$$

and therefore,

$$\tan(\theta_1 - \theta_2) = \frac{2r_1(x - x^{-1})\sin\theta_1}{2r^2 + 2 - 2r(x + x^{-1})\cos\theta} \qquad \text{by (S.2), (S.5)}$$

$$= \frac{r_1(x - x^{-1})\sin\theta_1}{r^2 + 1 - r(x + x^{-1})\cos\theta}$$

$$= \frac{r(x - x^{-1})\sin\theta}{r^2 + 1 - r(x + x^{-1})\cos\theta} \qquad \text{by (S.6)}$$

For $r = 1$ we obtain (5.68).

5.11 Using $\Gamma = \frac{\omega(z_0)\,dz_0}{2\pi}$, $\tilde{y} = \sqrt{z_0^2 - 1}\,\sqrt{1 - z^2}$, $\tilde{x} = (1 - z_0 z)$, we write

$$d\phi_{z_0}^+ = \Gamma \arctan 2\,[\tilde{y}, \tilde{x}] \qquad d\phi_{z_0}^+ = \Gamma \arctan 2\,[-\tilde{y}, \tilde{x}]$$

The transition $\tilde{x} + i\tilde{y} \mapsto \tilde{x} - i\tilde{y}$ is achieved by reflection over x-axis. Therefore, it holds for the corresponding angles that

$$\phi = \arctan2\,[\tilde{y}, \tilde{x}], \qquad -\phi = \arctan2\,[-\tilde{y}, \tilde{x}]$$

5.12 For the free vortex, we replace $x_0 = z_0$ and calculate the partial derivatives from (5.79) by (5.69):

$$\Delta p_{x_0} = -\frac{\varrho u_\infty\, \omega(x_0)\, \mathrm{d}x_0}{\pi} \frac{x_0 + x}{\sqrt{x_0^2 - 1}\,\sqrt{1 - x^2}} \qquad \text{for } -1 \leq x \leq 1,\ x_0 > 1$$

The formula describes the pressure difference that occurs at the wing point x due to the vortex pair with free element at x_0. The vortex pair causes the lift

$$l(x_0) = \int_{-1}^{1} \Delta p_{x_0}\, \mathrm{d}x = -\frac{\varrho u_\infty\, \omega(x_0)\, \mathrm{d}x_0}{\pi\sqrt{x_0^2 - 1}} \int_{-1}^{1} \frac{x_0 + x}{\sqrt{1 - x^2}}\, \mathrm{d}x$$

Because of its odd integrand, it is

$$\int_{-1}^{1} \frac{x}{\sqrt{1 - x^2}}\, \mathrm{d}x = 0$$

and consequently

$$\int_{-1}^{1} \frac{x_0 + x}{\sqrt{1 - x^2}}\, \mathrm{d}x = x_0 \int_{-1}^{1} \frac{\mathrm{d}x}{\sqrt{1 - x^2}} + \int_{-1}^{1} \frac{x}{\sqrt{1 - x^2}}\, \mathrm{d}x = \pi x_0$$

$$l(x_0) = -\frac{\varrho u_\infty\, \omega(x_0)\, \mathrm{d}x_0}{\pi\sqrt{x_0^2 - 1}} \int_{-1}^{1} \frac{x_0 + x}{\sqrt{1 - x^2}}\, \mathrm{d}x$$

$$= -\frac{\varrho u_\infty\, \omega(x_0)\, x_0\, \mathrm{d}x_0}{\sqrt{x_0^2 - 1}} \qquad \text{for } x_0 > 1$$

5.13 Using (4.10) and Stokes' theorem, we find that

$$\Gamma_C = \oint_C \mathbf{u}\, \mathrm{d}\mathbf{x} = \int_\Sigma \mathrm{rot}\,\mathbf{u} \cdot \mathbf{n}\, \mathrm{d}S = \int_\Sigma \boldsymbol{\omega} \cdot \mathbf{n}\, \mathrm{d}S$$

If the vortex tube can be reduced to a vortex filament due to cross-sectional constriction, we obtain its strength by

$$\lim_{|\Sigma| \to 0} \frac{\Gamma_C}{|\Sigma|} = \lim_{|\Sigma| \to 0} \frac{\int_\Sigma \boldsymbol{\omega} \cdot \mathbf{n}\, \mathrm{d}S}{|\Sigma|} = \omega$$

5.14 By the streamline curvature theorem (Theorem 4.1), there is a suction towards the vortex centre. This forms a vortex filament which, according to Helmholtz's theorem, does not end in the fluid. Instead, it extends from the surface to the bottom, where the flow velocity becomes zero. Close to the ground, there is a risk of becoming trapped in the obstacle. However, if you lie flat on the surface of the water, the smallest possible part of your body is in the area of the vortex filament and the suction effect is minimal.

To Chap. 6

6.1 In Fig. 6.6, right triangles with α_i are similar. Therefore, $\frac{L'}{L'_{\text{res}}} = \frac{u_\infty}{u_{\text{res}}}$. Rearranging by (6.10) yields

$$L' = L'_{\text{res}} \frac{u_\infty}{u_{\text{res}}} = \varrho\, u_{\text{res}}\, \Gamma \frac{u_\infty}{u_{\text{res}}} = \varrho\, u_\infty\, \Gamma$$

6.2 The similarity of the right triangles in Fig. 6.6 implies $\frac{D'_i}{L'_{\text{res}}} = \frac{w}{u_{\text{res}}}$. Rearranging by (6.10) yields

$$D'_i = L'_{\text{res}} \frac{w}{u_{\text{res}}} = \varrho\, u_{\text{res}} \Gamma \frac{w}{u_{\text{res}}} = \varrho w \Gamma$$

6.3 Use (6.9) with (6.8), (6.17), (6.18).

6.4 From (6.16) and (6.13) we get

$$C_L = \frac{L}{q_\infty S} = \frac{\varrho_\infty u_\infty \int_{-b/2}^{b/2} \Gamma(y)\, dy}{q_\infty S}$$

By substitution of $q_\infty = \frac{1}{2}\varrho_\infty u_\infty^2$ we obtain (6.22). For small α_i, we have $\sin \alpha_i \approx \alpha_i$. By (6.14) and (6.11) we get (6.23). From (6.17) and (6.23) we obtain (6.24).

6.5 From (6.13) and (6.28) we get

$$L = \varrho_\infty u_\infty \Gamma_0 \int_\pi^0 \sin \theta \, (-\frac{b}{2}) \sin \theta \, d\theta = \varrho_\infty u_\infty \Gamma_0 \frac{b}{2} \int_0^\pi \sin^2 \theta \, d\theta$$

We evaluate the integral using $\sin^2 \theta = \frac{1}{2}(1 - \cos(2\theta))$ and the subsequent variable substitution $\phi = 2\theta$.

$$\int_0^\pi \sin^2 \theta \, d\theta = \frac{\pi}{2}$$

This gives (6.29). By (6.16) we get (6.30).

Fig. A.8 Only the functions $\sin(n\theta)$ with odd n are symmetric around $\theta = \pi/2$.

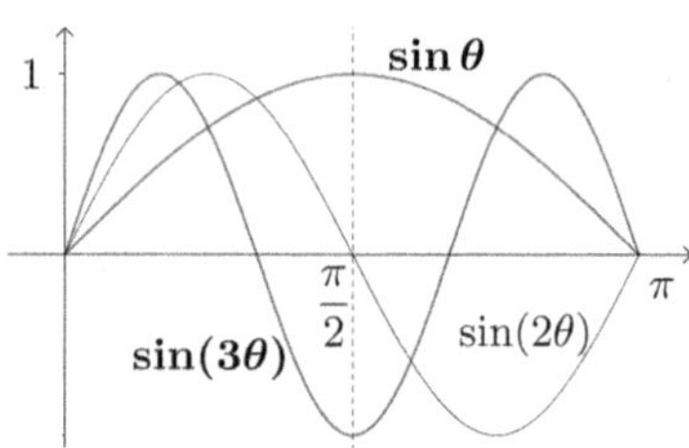

6.6 According to (6.28), elliptical wings have an elliptical circulation distribution $\Gamma(\theta) = \Gamma_0 \sin\theta$ ($0 \le \theta \le \pi$). In the horseshoe vortex, the vorticity $\frac{d\Gamma}{d\theta} = \Gamma_0 \cos\theta$ is released at the point $y = \frac{b}{2}\cos\theta$ of the lifting line. At the wing tips $y_1 = -\frac{b}{2}$ ($\theta_1 = \pi$) and $y_2 = \frac{b}{2}$ ($\theta_2 = 0$) we find the maximum of the released vorticity.

6.7 Using $\sin\alpha \sin\beta = \frac{1}{2}[\cos(\alpha - \beta) - \cos(\alpha + \beta)]$ for $\alpha = nx$, $\beta = mx$, we get

$$\sin(nx)\,\sin(mx) = \frac{1}{2}[\cos(n - m)x - \cos(n + m)x]$$

Case 1: For $m = n$ it follows that $\sin(nx)\,\sin(nx) = \frac{1}{2}[1 - \cos(2nx)]$. Thus,

$$\int_0^\pi \sin(nx)\,\sin(nx)\,dx = \frac{1}{2}\left(\int_0^\pi dx - \int_0^\pi \cos(2nx)\,dx\right) = \frac{\pi}{2}$$

Case 2: For $m \neq n$ it holds that

$$\int_0^\pi \sin(nx)\,\sin(mx)\,dx$$

$$= \frac{1}{2}\left(\int_0^\pi \cos(n - m)x\,dx - \int_0^\pi \cos(n + m)x\,dx\right)$$

$$= \frac{1}{2}\left(\frac{1}{n - m}\int_0^{(n-m)\pi} \cos z\,dz - \frac{1}{n + m}\int_0^{(n+m)\pi} \cos z\,dz\right)$$

$$= \frac{1}{2}\left(\frac{1}{n - m}\sin z\,\big|_0^{(n-m)\pi} - \frac{1}{n + m}\sin z\,\big|_0^{(n+m)\pi}\right) = 0$$

6.8 For symmetric circulations, only components $\sin(n\theta)$ are allowed that are symmetric around the centre line $y = 0$, i.e. $\theta = \frac{\pi}{2}$ (Fig. A.8).

They have a horizontal tangent at $\pi/2$. From $\frac{d}{d\theta}\sin(n\theta)\,\big|_{\theta=\frac{\pi}{2}} = 0$ it follows that $\cos(n\frac{\pi}{2}) = 0$. Thus, $n = 2k + 1$.

To Chap. 7

7.1 Let $F(\phi) = \frac{\phi}{2} r^2$ be the area of a circular sector. Then,

$$S = F(2\phi) = \phi r^2$$

Per unit of time, the air mass $m_i = S v_i \varrho = \phi r^2 v_i \varrho$ flows through S. Through the wing beat, it receives the impulse $\Delta I = m_i v_w = \phi r^2 \varrho v_i v_w$. Thus, it has the kinetic energy $P_i = \frac{1}{2} m_i v_w^2 = \frac{1}{2} \varrho \phi r^2 v_i v_w^2$ in the wake.

7.2 Equating (7.4) and (7.8) yields $v_w = 2v_i$. By replacing v_w in (7.8) we obtain

$$v_i = \sqrt{\frac{mg}{2\varrho r^2 \phi}} \tag{S.7}$$

If we substitute $v_w = 2v_i$ into (7.3) and use (S.7), we get

$$P_i = 2\varrho \phi r^2 v_i^3 = \sqrt{\frac{m^3 g^3}{2\varrho r^2 \phi}}$$

7.3 From

$$\frac{d\hat{w}}{dz} = \frac{dw}{d\zeta} \frac{d\zeta}{dz}, \quad \left(\frac{d\hat{w}}{dz}\right)^2 = \left(\frac{dw}{d\zeta}\right)^2 \left(\frac{d\zeta}{dz}\right)^2, \quad z\,dz = \frac{1}{2}\left(\zeta + \frac{1}{\zeta}\right) \frac{d\zeta}{dz}\,d\zeta$$

we obtain

$$\left(\frac{d\hat{w}}{dz}\right)^2 z\,dz = \left(\frac{dw}{d\zeta}\right)^2 \frac{d\zeta}{dz} \frac{1}{2}\left(\zeta + \frac{1}{\zeta}\right) d\zeta \tag{S.8}$$

Using

$$\frac{dz}{d\zeta} = \frac{1}{2}\left(1 - \frac{1}{\zeta^2}\right), \qquad \frac{d\zeta}{dz} = \frac{1}{dz/d\zeta} = \frac{2}{1 - \frac{1}{\zeta^2}}$$

it follows from (S.8) that

$$\left(\frac{d\hat{w}}{dz}\right)^2 z\,dz = \left(\frac{dw}{d\zeta}\right)^2 \left(\zeta + \frac{1}{\zeta}\right)\left(1 - \frac{1}{\zeta^2}\right)^{-1} d\zeta \tag{S.9}$$

From the series expansion

$$(1-x)^{-1} = 1 + x + x^2 + x^3 + \dots \qquad \text{for } |x| < 1$$

we find that

$$\left(1 - \frac{1}{\zeta^2}\right)^{-1} = 1 + \frac{1}{\zeta^2} + \frac{1}{\zeta^4} + \frac{1}{\zeta^6} + \dots \qquad \text{for } |\zeta| > 1$$

and thus

$$\left(\zeta + \frac{1}{\zeta}\right)\left(1 - \frac{1}{\zeta^2}\right)^{-1} = \zeta + \frac{1}{\zeta} + \frac{1}{\zeta^3} + \frac{1}{\zeta^5} + \dots + \frac{1}{\zeta} + \frac{1}{\zeta^3} + \frac{1}{\zeta^5} + \dots$$

$$= \zeta + \frac{2}{\zeta} + \frac{2}{\zeta^3} + \frac{2}{\zeta^5} + \dots \tag{S.10}$$

From (S.9) and (S.10) we obtain (7.17).

7.4 From (6.18) we obtain

$$D = \frac{1}{2}\varrho_\infty u_\infty^2 S C_D = \frac{1}{2}\varrho_\infty u_\infty^2 S C_d + \frac{1}{2}\varrho_\infty u_\infty^2 S C_{D_i} \quad \text{by (6.19)}$$

$$= \frac{1}{2}\varrho_\infty u_\infty^2 S C_d + \frac{1}{2}\varrho_\infty u_\infty^2 S \frac{C_L^2}{\pi e \mathit{AR}} \quad \text{by (6.53)}$$

$$= \frac{1}{2}\varrho_\infty u_\infty^2 S C_d + \frac{2L^2}{\varrho_\infty u_\infty^2 S} \frac{1}{\pi e \mathit{AR}} \quad \text{by (6.16)}$$

$$= \frac{1}{2}\varrho_\infty u_\infty^2 S C_d + \frac{2m^2 g^2}{\varrho_\infty u_\infty^2 S} \frac{1}{\pi e \mathit{AR}} \quad \text{by (7.21)}$$

7.5 Using (6.9), (6.16), (6.18), (6.19) we get

$$D = L\frac{D}{L} = mg\frac{D}{L} = mg\frac{C_D}{C_L} = mg\frac{C_d + C_{D_i}}{C_L}$$

Using (6.53) and the Oswald efficiency number e, we obtain (7.27).

To Chap. 8

8.1 Along a trajectory on the surface,

$$\frac{D}{Dt}z = \frac{D}{Dt}\eta(x, t)$$

$$\frac{\partial z}{\partial t} + \mathbf{u} \cdot \nabla z = \frac{\partial \eta}{\partial t} + \mathbf{u} \cdot \nabla \eta$$

The Cartesian coordinates x, y, z are both independent of each other and independent of t. Then, $\frac{\partial z}{\partial x} = \frac{\partial z}{\partial y} = \frac{\partial z}{\partial t} = 0$, $\frac{\partial z}{\partial z} = 1$. Using $\mathbf{u} = (u, 0, w)$, it follows that

$$w = \frac{\partial \eta}{\partial t} + u \frac{\partial \eta}{\partial x} + w \frac{\partial \eta}{\partial z} \qquad \text{because of } \frac{\partial z}{\partial t} = 0, \ \nabla z = (0, 0, 1)$$

$$w = \frac{\partial \eta}{\partial t} + u \frac{\partial \eta}{\partial x} \qquad \text{because } \eta(x, t) \text{ is independent of } z$$

8.2 From (8.22) we get $A'(z) = kC_1 \sinh(kz + C_2)$. (8.16) implies

$$\frac{\partial \phi}{\partial z} \Big|_{z=-h} = A'(-h) \sin(\omega t - kx + \phi_0) = 0$$

Thus, $A'(-h) = 0$. This leads to (8.23).

8.3 From (8.2), (8.20), (8.21) and (8.28) we get

$$c = \frac{\lambda}{T} = \frac{2\pi/k}{2\pi/\omega} = \frac{1}{k}\sqrt{gk \tanh(hk)}$$

8.4

(a) Let $u = (\mathbf{u} \cdot \mathbf{u})^{1/2}$. We parametrise C by arc length s. From

$$\mathbf{u} \cdot d\mathbf{u} = \mathbf{u} \cdot \frac{d\mathbf{u}}{ds} \, ds = \frac{1}{2} \frac{d(u^2)}{ds} \, ds$$

it follows that $\oint_C \mathbf{u} \, d\mathbf{u} = \frac{1}{2} \oint_C \frac{d(u^2)}{ds} \, ds = 0$.

(b) By (4.10), the rate of change of circulation is given by

$$\frac{d\Gamma}{dt} = \frac{d}{dt} \oint_C \mathbf{u} \, d\mathbf{x} = \oint_C \frac{d\mathbf{u}}{dt} \, d\mathbf{x} + \oint_C \mathbf{u} \, d(\frac{d\mathbf{x}}{dt})$$

$$= \oint_C \frac{d\mathbf{u}}{dt} \, d\mathbf{x} + \oint_C \mathbf{u} \, d\mathbf{u} = \oint_C \frac{d\mathbf{u}}{dt} \, d\mathbf{x} = \oint_C \mathbf{a} \, d\mathbf{x}$$

8.5 We use the following terms.

- r original radius, dV/dt original volumetric flow rate
- $r_1 = 0, 8r$ reduced radius, dV_1/dt reduced volumetric flow rate

From (8.36) we obtain the formula $dV/dt = cr^4$ with a constant c. Then, $dV_1/dt = c(0.8\,r)^4$ and thus

$$\frac{dV_1/dt}{dV/dt} = \frac{(0.8\,r)^4}{r^4} = 0.8^4 \approx 0.4$$

The volumetric flow rate decreases by 60%.

To "Formulas and Tables"

A.1 If we apply l_{kj} on (A.5), we get

$$l_{kj}\bar{x}_j = l_{kj}l_{ij}x_i$$

$$l_{kj}\bar{x}_j = \delta_{ki}x_i \qquad \text{by (A.4)}$$

$$l_{kj}\bar{x}_j = x_k$$

A.2 From (A.7) we get $l_{ip}l_{jq}\delta_{ij} = l_{jp}l_{jq} = \delta_{qp}$. Here we used (A.4) in the last step.

A.3 To prove the isotropy of $(\delta_{ij}\delta_{km})$ we use the approach (A.7).

$$l_{ip}l_{jq}l_{kr}l_{ms}\delta_{ij}\delta_{km} = l_{jp}l_{jq}l_{kr}l_{ks} = \delta_{pq}\delta_{rs}$$

In the last step, we used (A.4) again. For $(\delta_{ik}\delta_{jm})$ or $(\delta_{im}\delta_{jk})$, the proof is analogous. Thus, A is isotropic as a linear combination of isotropic tensors.

A.4 By s^{-1} we denote the inverse to (A.10). Then, the parametrisation of the curve by arc length can be written as a concatenation $\mathbf{x}(s) = \hat{\mathbf{x}} \circ s^{-1}$. Using the chain rule, we form the derivative.

$$\mathbf{x}'(s) = \hat{\mathbf{x}}'(s^{-1}) \frac{1}{ds/dt} = \frac{\hat{\mathbf{x}}'(t)}{\|\hat{\mathbf{x}}'(t)\|}$$

(To derive $s^{-1} = t(s)$, we have applied the derivative rule of the inverse function.) Thus, $\|\mathbf{x}'(s)\| = 1$.

A.5 Since $\mathbf{b}'$ is orthogonal to $\mathbf{b}$ (as is $\mathbf{t}'$ to $\mathbf{t}$ and $\mathbf{n}'$ to $\mathbf{n}$), $\mathbf{b}'$ lies in the common plane of $\mathbf{t}$ and $\mathbf{n}$. By differentiating (A.13) we get

$$\mathbf{b}' = \mathbf{t}' \times \mathbf{n} + \mathbf{t} \times \mathbf{n}'$$

$$= \mathbf{t} \times \mathbf{n}' \qquad \text{by (A.12)}$$

So $\mathbf{b}'$ is not only orthogonal to $\mathbf{b}$, but also to $\mathbf{t}$. Thus, $\mathbf{b}'$ is collinear to $\mathbf{n}$.

A.6 Because $i = e^{i\pi/2}$, we get

$$\frac{z}{i} = \frac{re^{i\phi}}{e^{i\frac{\pi}{2}}} = re^{i(\phi - \frac{\pi}{2})} \qquad \text{for } z = re^{i\phi}$$

Consequently, we obtain z/i from z by rotating by $-\pi/2$.

A.7

(a) $\bar{f}(z) = \overline{3\bar{z}^2 - 5i\bar{z}^3} = 3z + 5iz^3$

(b)   Using $z = re^{i\theta}$, $\bar{z} = re^{-i\theta}$, we get $f(z) = \ln r + i\theta$, $f(\bar{z}) = \ln r - i\theta$. Therefore,

$$\bar{f}(z) = \overline{f(\bar{z})} = f(z)$$

A.8 Let R be the radius of C. Using the parametrisation

$$z = \lambda(t) = Re^{it}, \qquad 0 \le t \le 2\pi,$$

we get $z^{-k} = R^{-k}e^{-ikt}$, $\lambda'(t) = iRe^{it}$. Therefore,

$$\oint_C \frac{dz}{z^k} = iR^{-k+1} \int_0^{2\pi} e^{-i(k-1)t} \, dt$$

For $k = 1$ we obtain $\oint_C \frac{dz}{z^k} = i \int_0^{2\pi} dt = 2\pi i$. For $k \ne 1$ we get

$$\oint_C \frac{dz}{z^k} = iR^{-k+1} \frac{e^{-i(k-1)t}}{-i(k-1)} \Big|_0^{2\pi} = 0$$

A.9 Let $z = z(t)$ be a smooth curve with $z(0) = z_0$. Then,

$$\frac{d}{dt} f(z(t)) \big|_{t=0} = \frac{df}{dz} \big|_{z=z_0} \frac{dz}{dt} \big|_{t=0}$$

$$\text{Arg} \frac{d}{dt} f(z(t)) \big|_{t=0} = \text{Arg} \frac{df}{dz} \big|_{z=z_0} + \text{Arg} \frac{dz}{dt} \big|_{t=0}$$

Consequently, we obtain the tangent of $f(z(t))$ at $f(z_0)$ from the tangent of $z(t)$ at z_0 by rotating it by an angle that is independent of the curve. Therefore, the intersection angle between two curves at z_0 remains constant after mapping by f.

A.10

(a) We multiply (A.23) by Γ_2 and (A.24) by Γ_1. Then we add the equations.

$$\Gamma_1 \dot{z}_1 + \Gamma_2 \dot{z}_2 = \frac{-i\Gamma_1\Gamma_2}{2\pi(\bar{z}_2 - \bar{z}_1)} + \frac{-i\Gamma_2\Gamma_1}{2\pi(\bar{z}_1 - \bar{z}_2)} = 0$$

(b) We differentiate (A.25) with respect to time and use (A.26).

$$\dot{z}_0 = \frac{\Gamma_1 \dot{z}_1 + \Gamma_2 \dot{z}_2}{\Gamma_1 + \Gamma_2} = \frac{\Gamma_1 \dot{z}_1 - \Gamma_1 \dot{z}_1}{\Gamma_1 + \Gamma_2} = 0$$

(c) Using Eqs. (A.23) and (A.24), we get

$$\frac{d}{dt}(z_1 - z_2) = \frac{-i\Gamma_2}{2\pi(\bar{z}_1 - \bar{z}_2)} - \frac{-i\Gamma_1}{2\pi(\bar{z}_2 - \bar{z}_1)} = -i\frac{\Gamma_1 + \Gamma_2}{2\pi(\bar{z}_1 - \bar{z}_2)}$$

$$(\bar{z}_1 - \bar{z}_2)\frac{d}{dt}(z_1 - z_2) = -i(\Gamma_1 + \Gamma_2)$$

$$(z_1 - z_2)\frac{d}{dt}(\bar{z}_1 - \bar{z}_2) = i(\Gamma_1 + \Gamma_2) \quad \text{by complex conjugation}$$

If we add both equations, we get

$$\overline{(z_1 - z_2)}\frac{d}{dt}(z_1 - z_2) + (z_1 - z_2)\frac{d}{dt}\overline{(z_1 - z_2)} = 0$$

$$\frac{d}{dt}[(z_1 - z_2)\overline{(z_1 - z_2)}] = 0$$

By $(z_1 - z_2)\overline{(z_1 - z_2)} = |z_1 - z_2|$ we obtain the assertion.

(d) The vortex center z_1 rotates at speed $|\dot{z}_1| = \omega\,|z_0 - z_1|$ around z_0. Thus,

$$\omega = \frac{|\dot{z}_1|}{|z_0 - z_1|}$$

For the numerator, we get from (A.24)

$$|\dot{z}_1| = \frac{\Gamma_2}{2\pi\,|z_1 - z_2|}$$

Using (A.25), it follows for the denominator that

$$|z_0 - z_1| = \left|\frac{\Gamma_1 z_1 + \Gamma_2 z_2}{\Gamma_1 + \Gamma_2} - z_1\right| = \frac{\Gamma_2\,|z_2 - z_1|}{\Gamma_1 + \Gamma_2}$$

In summary, we have

$$\omega = \frac{\Gamma_2}{2\pi\,|z_1 - z_2|}\bigg/\frac{\Gamma_2\,|z_2 - z_1|}{\Gamma_1 + \Gamma_2} = \frac{\Gamma_1 + \Gamma_2}{2\pi\,|z_2 - z_1|^2}$$

A.11 For a scalar function ϕ, it holds $d\phi = \nabla\phi \cdot \mathbf{dr}$. By substitution of

$$d\phi = \sum_n \frac{\partial\phi}{\partial\xi_n}d\xi_n, \quad \nabla\phi = \sum_n (\nabla\phi)_n \mathbf{e}_n, \quad \mathbf{dr} = \sum_n h_n \mathbf{e}_n d\xi_n,$$

we get $\sum_n \frac{\partial \phi}{\partial \xi_n} d\xi_n = \sum_n (\nabla \phi)_n h_n d\xi_n$. A coefficient comparison yields

$$\frac{\partial \phi}{\partial \xi_n} = (\nabla \phi)_n h_n$$

$$(\nabla \phi)_n = \frac{1}{h_n} \frac{\partial \phi}{\partial \xi_n}$$

Since $\frac{\partial \xi_k}{\partial \xi_n} = \delta_{kn}$, we find that $\nabla \xi_k = \sum_n \frac{1}{h_n} \mathbf{e}_n \frac{\partial \xi_k}{\partial \xi_n} = \frac{1}{h_k} \mathbf{e}_k$.

A.12 It holds that

$$0 = \nabla \times (\nabla \xi_k) \qquad \text{by (1.19)}$$

$$0 = \nabla \times \left(\frac{1}{h_k} \mathbf{e}_k\right) \qquad \text{by (A.36)}$$

$$0 = \left(\nabla \frac{1}{h_k}\right) \times \mathbf{e}_k + \frac{1}{h_k} (\nabla \times \mathbf{e}_k) \qquad \text{by (1.13)}$$

Thus, $\frac{1}{h_k}(\nabla \times \mathbf{e}_k) = -(\nabla \frac{1}{h_k}) \times \mathbf{e}_k = \mathbf{e}_k \times (\nabla \frac{1}{h_k})$. Multiplication by h_k yields the assertion.

A.13 By (A.35),

$$\nabla\left(\frac{1}{h_1}\right) = \frac{1}{h_1} \frac{\partial(1/h_1)}{\partial \xi_1} \mathbf{e}_1 + \frac{1}{h_2} \frac{\partial(1/h_1)}{\partial \xi_2} \mathbf{e}_2 + \frac{1}{h_3} \frac{\partial(1/h_1)}{\partial \xi_3} \mathbf{e}_3$$

$$\nabla\left(\frac{1}{h_1}\right) = -\frac{1}{h_1^3} \frac{\partial h_1}{\partial \xi_1} \mathbf{e}_1 - \frac{1}{h_1^2 h_2} \frac{\partial h_1}{\partial \xi_2} \mathbf{e}_2 - \frac{1}{h_1^2 h_3} \frac{\partial h_1}{\partial \xi_3} \mathbf{e}_3$$

$$h_1 \mathbf{e}_1 \times \nabla\left(\frac{1}{h_1}\right) = -\frac{1}{h_1 h_2} \frac{\partial h_1}{\partial \xi_2} \mathbf{e}_3 + \frac{1}{h_1 h_3} \frac{\partial h_1}{\partial \xi_3} \mathbf{e}_2$$

Using (A.37), we get

$$\nabla \times \mathbf{e}_1 = h_1\left(-\frac{1}{h_1^2 h_2} \frac{\partial h_1}{\partial \xi_2} \mathbf{e}_3 + \frac{1}{h_1^2 h_3} \frac{\partial h_1}{\partial \xi_3} \mathbf{e}_2\right)$$

$$= -\frac{1}{h_1 h_2} \frac{\partial h_1}{\partial \xi_2} \mathbf{e}_3 + \frac{1}{h_1 h_3} \frac{\partial h_1}{\partial \xi_3} \mathbf{e}_2$$

The rest follows in the same manner.

A.14 It is

$$\nabla e_1 = \nabla(e_2 \times e_3) \; = \; e_3(\nabla \times e_2) - e_2(\nabla \times e_3)$$

$$= \frac{1}{h_1 h_2} \frac{\partial h_2}{\partial \xi_1} + \frac{1}{h_1 h_3} \frac{\partial h_3}{\partial \xi_1} \qquad \text{by (A.38)}$$

$$= \frac{1}{h_1 h_2 h_3}(h_3 \frac{\partial h_2}{\partial \xi_1} + h_2 \frac{\partial h_3}{\partial \xi_1}) \; = \; \frac{1}{h_1 h_2 h_3} \frac{\partial (h_2 h_3)}{\partial \xi_1}$$

The rest follows in the same manner.

A.15 From (1.15) it follows that $\operatorname{div} \mathbf{q} = \sum_k (q_k \nabla e_k + e_k \cdot \nabla q_k)$. Using (A.35), we evaluate the right-hand side

$$\sum_k e_k \cdot \nabla q_k = \sum_k e_k \cdot \left[\frac{1}{h_1} \frac{\partial q_k}{\partial \xi_1} e_1 + \frac{1}{h_2} \frac{\partial q_k}{\partial \xi_2} e_2 + \frac{1}{h_3} \frac{\partial q_k}{\partial \xi_3} e_3 \right]$$

$$= \frac{1}{h_1 h_2 h_3} \left[h_2 h_3 \frac{\partial q_1}{\partial \xi_1} + h_1 h_3 \frac{\partial q_2}{\partial \xi_2} + h_1 h_2 \frac{\partial q_3}{\partial \xi_3} \right]$$

as well as by (A.39)

$$\sum_k q_k \nabla e_k = \frac{1}{h_1 h_2 h_3} \left[q_1 \frac{\partial (h_2 h_3)}{\partial \xi_1} + q_2 \frac{\partial (h_1 h_3)}{\partial \xi_2} + q_3 \frac{\partial (h_1 h_2)}{\partial \xi_3} \right]$$

Therefore,

$$\operatorname{div} \mathbf{q} = \frac{1}{h_1 h_2 h_3} \left[q_1 \frac{\partial (h_2 h_3)}{\partial \xi_1} + h_2 h_3 \frac{\partial q_1}{\partial \xi_1} + q_2 \frac{\partial (h_1 h_3)}{\partial \xi_2} + h_1 h_3 \frac{\partial q_2}{\partial \xi_2} \right.$$

$$\left. + q_3 \frac{\partial (h_1 h_2)}{\partial \xi_3} + h_1 h_2 \frac{\partial q_3}{\partial \xi_3} \right]$$

$$= \frac{1}{h_1 h_2 h_3} \left[\frac{\partial (q_1 h_2 h_3)}{\partial \xi_1} + \frac{\partial (q_2 h_1 h_3)}{\partial \xi_2} + \frac{\partial (q_3 h_1 h_2)}{\partial \xi_3} \right]$$

A.16 Let us consider $\Delta \phi = \operatorname{div} \operatorname{grad} \phi$, where we calculate the components of the gradient $(\nabla \phi)_k = \frac{1}{h_k} \frac{\partial \phi}{\partial \xi_k}$ by (A.35). Substituting $q_k = \frac{1}{h_k} \frac{\partial \phi}{\partial \xi_k}$ into (A.40) gives the result.

A.17 It holds that

$$
\frac{D\mathbf{u}}{Dt} = \frac{\partial \mathbf{u}}{\partial t} + \frac{d\mathbf{r}}{dt}\frac{\partial \mathbf{u}}{\partial \mathbf{r}}
$$

$$
= \frac{\partial \mathbf{u}}{\partial t} + \frac{dr}{dt}\frac{\partial \mathbf{u}}{\partial r} + \frac{d\theta}{dt}\frac{\partial \mathbf{u}}{\partial \theta} + \frac{dz}{dt}\frac{\partial \mathbf{u}}{\partial z}
$$

$$
= \frac{\partial \mathbf{u}}{\partial t} + u_r\frac{\partial \mathbf{u}}{\partial r} + \frac{u_\theta}{r}\frac{\partial \mathbf{u}}{\partial \theta} + u_z\frac{\partial \mathbf{u}}{\partial z} \tag{S.11}
$$

Recall that the derivative $\partial \mathbf{u}/\partial t$ is calculated at a fixed location. Thus, the unit vectors remain fixed and their time derivatives vanish. It follows that

$$
\frac{\partial \mathbf{u}}{\partial t} = \dot{u}_r\mathbf{e}_r + \dot{u}_\theta\mathbf{e}_\theta + \dot{u}_z\mathbf{e}_z \tag{S.12}
$$

Furthermore, it is

$$
\frac{\partial \mathbf{u}}{\partial r} = \frac{\partial}{\partial r}(u_r\mathbf{e}_r + u_\theta\mathbf{e}_\theta + u_z\mathbf{e}_z) = \frac{\partial u_r}{\partial r}\mathbf{e}_r + \frac{\partial u_\theta}{\partial r}\mathbf{e}_\theta + \frac{\partial u_z}{\partial r}\mathbf{e}_z \quad \text{by (A.43)}
$$

$$
\frac{\partial \mathbf{u}}{\partial \theta} = \frac{\partial}{\partial \theta}(u_r\mathbf{e}_r + u_\theta\mathbf{e}_\theta + u_z\mathbf{e}_z)
$$

$$
= \frac{\partial u_r}{\partial \theta}\mathbf{e}_r + \frac{\partial u_\theta}{\partial \theta}\mathbf{e}_\theta + \frac{\partial u_z}{\partial \theta}\mathbf{e}_z + u_r\mathbf{e}_\theta - u_\theta\mathbf{e}_r \quad \text{by (A.45)}
$$

$$
\frac{\partial \mathbf{u}}{\partial z} = \frac{\partial}{\partial z}(u_r\mathbf{e}_r + u_\theta\mathbf{e}_\theta + u_z\mathbf{e}_z) = \frac{\partial u_r}{\partial z}\mathbf{e}_r + \frac{\partial u_\theta}{\partial z}\mathbf{e}_\theta + \frac{\partial u_z}{\partial z}\mathbf{e}_z \quad \text{by (A.44)}
$$

From (S.11), (A.47), (S.12) we obtain the assertion.

A.18 According to (A.50), we calculate the components of $\Delta\mathbf{u}$.

$$
\frac{\partial^2 \mathbf{u}}{\partial r^2} = \frac{\partial^2 u_r}{\partial r^2}\mathbf{e}_r + \frac{\partial^2 u_\theta}{\partial r^2}\mathbf{e}_\theta + \frac{\partial^2 u_z}{\partial r^2}\mathbf{e}_z \quad \text{by (A.43)}
$$

$$
\frac{1}{r}\frac{\partial \mathbf{u}}{\partial r} = \frac{1}{r}\frac{\partial u_r}{\partial r}\mathbf{e}_r + \frac{1}{r}\frac{\partial u_\theta}{\partial r}\mathbf{e}_\theta + \frac{1}{r}\frac{\partial u_z}{\partial r}\mathbf{e}_z \quad \text{by (A.43)}
$$

$$
\frac{1}{r^2}\frac{\partial^2 \mathbf{u}}{\partial \theta^2} = \frac{1}{r^2}\left[\frac{\partial^2 u_r}{\partial \theta^2} - 2\frac{\partial u_\theta}{\partial \theta} - u_r\right]\mathbf{e}_r + \frac{1}{r^2}\left[\frac{\partial^2 u_\theta}{\partial \theta^2} + 2\frac{\partial u_r}{\partial \theta} - u_r\right]\mathbf{e}_\theta
$$

$$
+ \frac{1}{r^2}\frac{\partial^2 u_z}{\partial \theta^2}\mathbf{e}_z \qquad\qquad \text{by (A.45)}
$$

$$
\frac{\partial^2 \mathbf{u}}{\partial z^2} = \frac{\partial^2 u_r}{\partial z^2}\mathbf{e}_r + \frac{\partial^2 u_\theta}{\partial z^2}\mathbf{e}_\theta + \frac{\partial^2 u_z}{\partial z^2}\mathbf{e}_z \qquad \text{by (A.44)}
$$

In summary, we have

$$
\Delta\mathbf{u} = \left[\frac{\partial^2 u_r}{\partial r^2} + \frac{1}{r}\frac{\partial u_r}{\partial r} + \frac{1}{r^2}\left(\frac{\partial^2 u_r}{\partial \theta^2} - 2\frac{\partial u_\theta}{\partial \theta} - u_r\right) + \frac{\partial^2 u_r}{\partial z^2}\right]\mathbf{e}_r
$$
$$
+ \left[\frac{\partial^2 u_\theta}{\partial r^2} + \frac{1}{r}\frac{\partial u_\theta}{\partial r} + \frac{1}{r^2}\left(\frac{\partial^2 u_\theta}{\partial \theta^2} + 2\frac{\partial u_r}{\partial \theta} - u_r\right) + \frac{\partial^2 u_\theta}{\partial z^2}\right]\mathbf{e}_\theta
$$
$$
+ \left[\frac{\partial^2 u_z}{\partial r^2} + \frac{1}{r}\frac{\partial u_z}{\partial r} + \frac{1}{r^2}\left(\frac{\partial^2 u_z}{\partial \theta^2}\right) + \frac{\partial^2 u_z}{\partial z^2}\right]\mathbf{e}_z
$$

By (A.50), we obtain the assertion.

Index